GEOLOGY OF THE APPALACHIAN TRAIL
IN PENNSYLVANIA

To AL & LINDA
from Pete

At long last
it has arrived.

*"The spectator, while standing on this rupic emi-
nence, has a commanding view of one of the
most variegated sceneries imaginable."*
                          I. Daniel Rupp, 1845

The Pinnacle (site 13), Berks County.

General Geology Report 74

GEOLOGY OF THE APPALACHIAN TRAIL IN PENNSYLVANIA

by J. Peter Wilshusen
Pennsylvania Geological Survey

Prepared in cooperation with
the Keystone Trails Association

PENNSYLVANIA GEOLOGICAL SURVEY
FOURTH SERIES
HARRISBURG

1983

ISBN: 0-8182-0020-0

ADDITIONAL COPIES
OF THIS PUBLICATION MAY BE PURCHASED FROM
STATE BOOK STORE, P.O. BOX 1365
HARRISBURG, PENNSYLVANIA 17125

PREFACE

Awareness of the natural history of a mountainous area through which a traveler hikes enhances the walking experience and gives it greater depth. The foundation of that natural history is the bedrock over which a trail passes and which provides a landscape of mountains and valleys.

In the eastern United States the Appalachian Trail follows the spine of the Appalachian Mountain chain from Maine to Georgia. This book discusses geology in the vicinity of the Appalachian Trail in Pennsylvania. From many sites of geologic interest along the Trail, 40 have been chosen as outdoor classrooms in which geologic features can best be seen. Through geologic sketches, cross sections, photographs, a full-color geologic map, and text you are introduced to this important aspect of natural history.

Whether you are a through hiker keeping a tightly planned schedule, a weekend backpacker, or a day hiker who may want to return to a favorite section of the Trail time after time, this book will provide a look inside the mountain under your feet on the Appalachian Trail in Pennsylvania.

ACKNOWLEDGEMENTS

Special thanks for completion of this publication go to the Keystone Trails Association for financial support and to Maurice Forrester, David Raphael, Malcolm White, and Boyd Sponaugle of that organization for encouragement during its preparation.

Assistance from Alan Geyer and other staff members of the Pennsylvania Geological Survey during manuscript review, and field assistance from summer geology student employees at the Survey, are greatly appreciated.

The knowledge and experience of fellow hiker Henry Knauber contributed a great deal to the author's understanding of coal mining history and the history of Saint Anthony's Wilderness in the vicinity of the Appalachian Trail.

CONTENTS

PLATE

(in pocket)

Plate 1. Geologic map of the Appalachian Trail and vicinity in Pennsylvania.

GEOLOGY OF THE APPALACHIAN TRAIL
IN PENNSYLVANIA

by

J. Peter Wilshusen

INTRODUCTION

The Appalachian Trail is a 2,097-mile continuous footpath along the crest of the Appalachian Mountains in the eastern United States. It passes through 14 states between its south end at Springer Mountain, Georgia, and its north end at Katahdin, Maine. Initially proposed in 1921 by Benton MacKaye, a forester and regional planner from Mas-

The longest continuously marked footpath in the world, the Appalachian Trail follows the spine of the Appalachian Mountains for 2,097 miles through 14 states in the eastern United States.

sachusetts, the Trail was completed in 1937, but was almost totally destroyed by the 1938 hurricane. After a halt in Trail activity during World War II, it was reopened and at last completed as a continuous path in 1951.

Pennsylvania is the keystone state of the Trail, the center of which is within this state's 228-mile segment. That segment extends across the southeastern part of Pennsylvania, from the Delaware Water Gap at the New Jersey State line to Pen Mar on the Maryland border. The Trail passes through three different topographic and geologic settings in this 228-mile traverse. Each one of these regions—(1) the Appalachian Mountain section of the Valley and Ridge physiographic province, (2) the Great Valley section, also of the Valley and Ridge physiographic province, and (3) the South Mountain section of the Blue Ridge physiographic province—provides a characteristic terrain for the Trail, which to a large degree explains why it was routed through these parts of Pennsylvania.

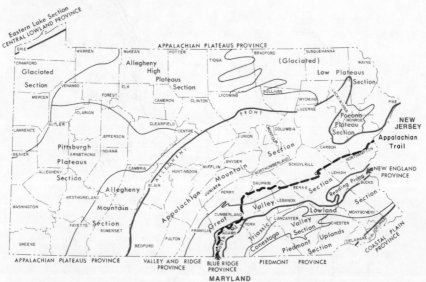

Physiographic provinces of Pennsylvania and the location of the Appalachian Trail.

THE GEOLOGIC SETTING

The Appalachian Mountains are an old and fairly stable mountain range in which mountain-building processes have been quiet for about 240 million years. Rocks in these mountains include types from all three classes: sedimentary, igneous, and metamorphic. Some of the rocks are more than 500 million years old.

An examination of the route of the Trail from north to south in Pennsylvania shows the following geologic features. From the Delaware Water Gap (Monroe County), the Trail follows a single ridge to Swatara Gap (Lebanon County) without change in the rock type. This ridge is called Kittatinny Mountain, and, to the south, Blue Mountain. The rock is 440-million-year-old quartzite and conglomerate of the Shawangunk and Tuscarora Formations (Plate 1). Kittatinny (Blue) Mountain is part of a range, extending north and west of the Trail, composed of sinuous ridges underlain by rock formations that were folded during periods of past geologic mountain building and eroded to the form they have today. In cross section these mountains are tight folds similar to wrinkles in a tablecloth that has been pushed across a table. Beyond these folded mountains to the north and west, set off by the Allegheny Front, are gently folded to flat-lying sedimentary rocks of the Appalachian Plateau. These relationships are shown in the adjacent schematic cross section, which has each physiographic section labeled.

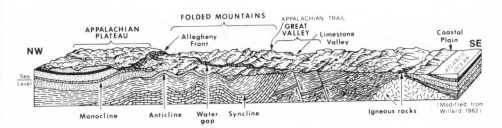

Schematic cross section through part of Pennsylvania showing the characteristic geologic structures in the different physiographic provinces.

Southeastward from the Trail a broad, open valley is underlain by intensely folded and faulted bedrock units of shale and limestone. These easily eroded and structurally deformed rocks are shown on the cross section as the Great Valley. Southeast from the Great Valley, forming its southeastern margin in this state, are ridges of igneous rocks and tough, resistant conglomerate and quartzite of the Reading Prong physiographic section. The Great Valley is oriented northeast-southwest in eastern Pennsylvania. However, it turns to the south as it passes out of Pennsylvania into Maryland and Virginia and is the same valley that is west of the Trail in those states.

At Swatara Gap (Lebanon County), the Trail turns toward the northwest, going more deeply into the folded mountains, following a route across sandstone, quartzite, conglomerate, and shale formations of the Appalachian Mountains (Plate 1). Up to this point, from the north, the route has been along the trend of the bedrock units; here, it goes across the trend in a zigzag route. It crosses a syncline at the southern end of the Pennsylvania anthracite district, then turns southwest to parallel the folded ridges again, continuing for several miles west of the Susquehanna River.

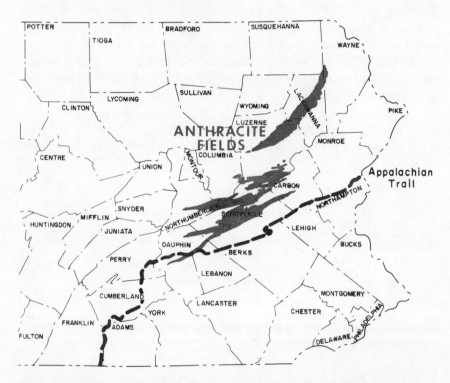

Index map showing the proximity of the Appalachian Trail to the anthracite coal region of Pennsylvania.

At the west end of Cove Mountain (Perry County), the Trail turns south, back across ridges of the folded mountains to the broad, open expanse of the Great Valley underlain by structurally deformed bedrock.

Across the Great Valley, the Trail ascends South Mountain (Cumberland and Franklin Counties) into the third distinct topographic and geologic setting in Pennsylvania. South Mountain is a complex faulted fold that has a core of ancient metamorphosed volcanic rocks surrounded by resistant quartzite rock units and by carbonate rocks in the adjacent valley to the north (see the cross section below). It is the northern tip of the massive Blue Ridge physiographic province, which becomes the dominant feature of the Appalachian Mountains to the south.

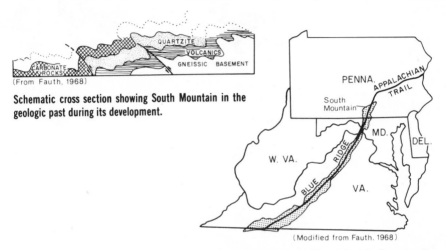

(From Fauth, 1968)

Schematic cross section showing South Mountain in the geologic past during its development.

(Modified from Fauth, 1968)

The Blue Ridge physiographic province in Pennsylvania and adjacent states.

The landscape in all three of the topographic and geologic settings in Pennsylvania results from the combined effects of rock type, structural deformation by folding and faulting, and erosion throughout geologic time. The landscape has also been affected by a relatively recent geologic event, glaciation, which occurred between 550,000 and 12,500 years ago and strongly modified the land surface in the vicinity of the Trail. The map *Glacial Deposits of Eastern Pennsylvania* shows the extent and age of three glacial advances and their proximity to the Trail corridor. Much of the route does not cross glaciated areas, but many geologic features along the route are the result of geologic processes that occurred near glacial margins. These features and processes are referred to as periglacial because they occurred close to the ice but not in direct contact with it. Rocks and ridges in this environment were profoundly affected by periglacial weathering. Outcrops exposed along crests of ridges were broken and massive blocks

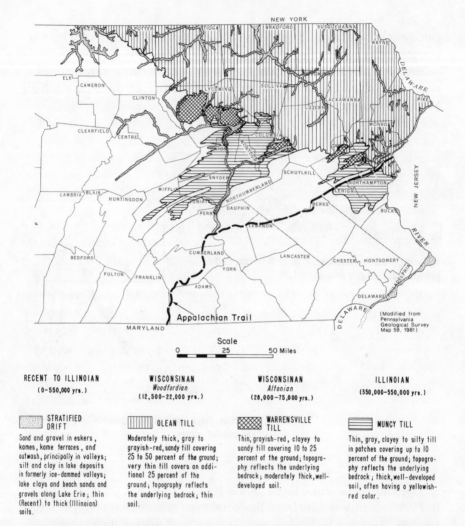

RECENT TO ILLINOIAN (0-550,000 yrs.)	WISCONSINAN *Woodfordian* (12,500-22,000 yrs.)	WISCONSINAN *Altonian* (28,000-75,000 yrs.)	ILLINOIAN (350,000-550,000 yrs.)
STRATIFIED DRIFT	OLEAN TILL	WARRENSVILLE TILL	MUNCY TILL
Sand and gravel in eskers, kames, kame terraces, and outwash, principally in valleys; silt and clay in lake deposits in formerly ice-dammed valleys; lake clays and beach sands and gravels along Lake Erie; thin (Recent) to thick (Illinoian) soils.	Moderately thick, gray to grayish-red, sandy till covering 25 to 50 percent of the ground; very thin till covers an additional 25 percent of the ground; topography reflects the underlying bedrock; thin soil.	Thin, grayish-red, clayey to sandy till covering 10 to 25 percent of the ground; topography reflects the underlying bedrock; moderately thick, well-developed soil.	Thin, gray, clayey to silty till in patches covering up to 10 percent of the ground; topography reflects the underlying bedrock; thick, well-developed soil, often having a yellowish-red color.

The heavy line on the southern part of a patterned area represents the limit of that glacial advance. Within the patterned areas, the tills occur as discontinuous deposits. The approximate percentage of each area that is actually covered by the till is stated in the descriptions above.

Glacial deposits of eastern Pennsylvania.

moved by frequently repeated freeze-thaw cycles that left broken, serrated, rocky-crested ridges which have boulder-strewn flanks and boulder accumulations on adjacent valley floors. At no time did the glacial ice advance farther south to smooth the rugged topography developed here, but melted back slowly northward to its present location near the Arctic Circle, leaving the effects of periglacial weathering behind.

THE TRAIL AND GEOLOGY

Writers have referred to the Trail in Pennsylvania as having a "gentle beauty." Hikers report it to be rocky; in fact, some say it is the rockiest stretch of the entire route. But, with all its rocks, the Trail in Pennsylvania is loved by many, hiked by thousands, and carefully maintained by a few.

The Trail is in a natural setting, and, as a result, many scenic and geologic features are seen by the hiker. These same features would not be visible to the motorist. In this sense the Appalachian Trail is the greatest of all nature trails.

The geology of the Trail is responsible for the landforms we see and the rocks we walk on. These rocks are composed of minerals, many of which are of economic importance in Pennsylvania. In some places plant and animal fossils are found in the rocks, and a clue to the geologic past may unfold. Thus, the geology of the Appalachian Trail in Pennsylvania is an interesting story. This publication has been written to tell that story. The regional geologic setting, the rocks along the Trail, the scenery, and outstanding geologic features are all described. Sketch maps, geologic cross sections, block diagrams, and photographs are used to illustrate the descriptions. The geologic map inside the back cover is a handy reference while reading this book. Locations of sites that are described in the book are shown on the map.

Throughout the discussions of the geology along the Trail, geologic terminology must be used. A glossary of geologic terms is provided in the back of the book. It is desirable to read this glossary first and become familiar with key words used in the text.

This publication does not include detailed explanations of many basic geologic concepts due to lack of space. For an explanation of these subjects, the reader is referred to books listed in the reference section.

GEOLOGIC SITES ALONG THE TRAIL

What and where are the outstanding geologic features along the Appalachian Trail? The following descriptions of 40 sites along 228 miles of the Trail will answer this question. Together they tell a story never before told. It is an interesting story; pause awhile and consider several or all of the sites as a group. They illustrate the diversity and beauty of Pennsylvania's geology and scenery.

Please note that the listing of a site off the Trail does not constitute permission to enter the property; always ask permission. Where the Trail is on state-owned land, do not carry away any natural specimens.

1. Delaware Water Gap
2. Wolf Rocks
3. Wind Gap
4. Smith Gap
5. Little Gap and Devils Potato Patch
6. Lehigh Gap and Devils Pulpit
7. Lehigh Furnace Gap
8. Bake Oven Knob
9. Bears Rocks
10. The Cliffs
11. Dans Pulpit
12. River of Rocks and Hawk Mountain
13. The Pinnacle
14. Pulpit Rock and Blue Rocks
15. Auburn Lookout
16. Schuberts Gap
17. Round Head
18. The Kessel
19. Monroe Valley Overlook
20. Swatara Gap
21. Rausch Gap
22. Strip Mine
23. Yellow Springs
24. DeHart Reservoir Overlook and Devils Race Course
25. Crest of Peters Mountain
26. Powells Valley Overlook
27. Susquehanna River Overlook
28. Hawk Rock
29. Cove Mountain
30. Great Valley
31. White Rocks
32. Pole Steeple
33. Pine Grove Furnace
34. Sunset Rocks
35. Lewis Rocks
36. Big Flat Tower
37. Quarry Gap and Caledonia Park
38. Snowy Mountain Tower
39. Chimney Rocks and Monument (Shaffers) Rock
40. Buzzards Roost

1. DELAWARE WATER GAP

The Delaware Water Gap, the northern terminus of the Appalachian Trail in Pennsylvania, is a feature of national geologic significance. It is one of the best examples of a water gap in the United States.

The flowing water of the Delaware River has cut through erosion-resistant rock formations of Kittatinny Mountain to form the gap. This geologic action is related to a network of rivers and streams in the eastern United States that began to flow long ago on an upland surface (the landward extension of the coastal plain) before erosion carved out the present valleys and ridges. In time the rivers, which originated across the trend of the folded formations, cut the ridges, creating the water gap. At locations of most water gaps there is a weakness in the mountain due to a fault, flexure, or change in rock material, causing a ridge to be somewhat more easily eroded than in adjacent parts of the mountain.

The gap is cut through three rock formations: dark-gray shale (Martinsburg Formation, Ordovician age), quartzite (Shawangunk Formation, Ordovician and Silurian age), and red sandstone (Bloomsburg Formation, Silurian age). See the cross section below.

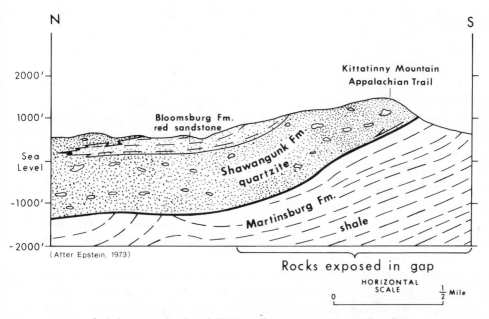

Geologic cross section through Kittatinny Mountain at the Delaware Water Gap.

The Delaware River, flowing from the north (top of photo), cuts through Kittatinny Mountain to form a spectacular water gap.

The dark-gray shale formation occupies the south end of the water gap. Some shale in the gap is covered by broken boulders and talus from the mountain above. The quartzite formation is the main rock layer visible in the gap and cliffs. The red sandstone formation under-lies the north end. Excellent exposures of this red sandstone are found at numerous places along the Trail about halfway up Mount Minsi.

Continental ice sheets extended into northeastern Pennsylvania during a period of geologic time called the Pleistocene. They spread into land now occupied by the Trail from the Delaware Water Gap southwest for about 35 miles into Lehigh County. The rest of the Trail in Pennsylvania has not been glaciated.

Three periods of glaciation occurred, named for the areas of the country where they were originally found and described. The oldest is the Illinoian glaciation, which occurred 550,000 to 350,000 years ago; the youngest are two Wisconsinan glaciations which took place 75,000 to 12,500 years ago. During this Great Ice Age the successive glaciations sculpted the landscape. Bedrock was scratched and grooved by advancing ice and till, composed of gravel, sand, silt, and clay, which was deposited as hills and terraces by retreating ice. Most glacial deposits and grooved bedrock surfaces in the water gap area resulted from Wisconsinan glaciation. At its line of maximum advance south of the gap, a terminal moraine of poorly sorted, jumbled rocks and sediment was deposited. As the glacier receded, meltwater from it ponded behind the moraine to form an 8-mile-long, 200-foot-deep lake north of Godfrey Ridge, called glacial Lake Sciota. The original outlet for the lake was in the terminal moraine at Saylorsburg. As the ice front retreated northeastward, Delaware Water Gap was uncovered and the lake drained through it.

The thickness of the glacial ice here during the Pleistocene is not known. Early geologic investigators speculated that it rose high above Kittatinny (Blue) Mountain, and that there were waterfalls and plunge pools at the ice front. There is no evidence to support this, but glacial grooves and striations on the top of 1,200-foot-high Kittatinny Mountain indicate that the water gap was filled and the mountains over-

Southbound on the Appalachian Trail in Pennsylvania, near the base of Mount Minsi, there is a view of the water gap and north-dipping sandstone and conglomerate on the flank of Mount Tammany.

Glacial scratches on a rock surface along the Appalachian Trail south of Lake Lenape. These striations indicate that movement of the ice at this point was in a direction parallel to the hammer handle.

topped by ice. Also, the variations in direction of glacial grooves and striations in this area indicate that preglacial topography had an influence on the direction of glacial ice movement.

On the east side of the Delaware River north of the water gap, old copper mines that date to the 1650's represent some of the earliest mining in the United States. Southwest of the gap, in the vicinity of Wind Gap and Lehigh Gap, minor occurrences of copper are present in red shales and sandstones north of the Trail. Nearby, Old Mine Road, which was the first long road to come into this area from the east in about 1659, reflects this past mining history.

At Totts Gap, 2.5 miles southwest of Delaware Water Gap, there is a short, abandoned mine opening. It is rumored that this was part of a gold-mine promotion deal in which stock was sold after gold-bearing rock was assayed. To make the assay impressive, gold dust was loaded in a shotgun shell and fired at the mine face. Samples from this face were then collected and assayed.

2. WOLF ROCKS

Wolf Rocks is an outcrop of quartzite (Shawangunk Formation) that forms the crest of The Little Offset ridge. This quartzite is medium gray, medium to coarse grained, and crossbedded, and contains some large quartz pebbles. It crops out here in a narrow band on the ridge as a short, broken stretch of bare rock forming a steep slope to the north and a relatively level hilltop to the south. The rock layers dip to the southeast on the north side of a syncline. The relationship of Wolf Rocks to the syncline is shown on the geologic map and cross section.

Wolf Rocks is significant because it marks the southernmost point of continental glaciation along the Trail. The effects of a glacial climate extend far to the south, but the ice stopped here.

The Wisconsinan ice advance left Pennsylvania about 12,500 years ago. However, evidence of it appears as though the glacier were with us very recently. A terminal moraine of boulders, gravel, silt, and sand carried forward by advancing ice and meltwater surrounds The Big and Little Offset ridges on the north and east. The Appalachian Trail

View looking east on the Appalachian Trail at Wolf Rocks on The Little Offset ridge. The outcrops are quartzite. Notice the dip of the rock layers to the southeast.

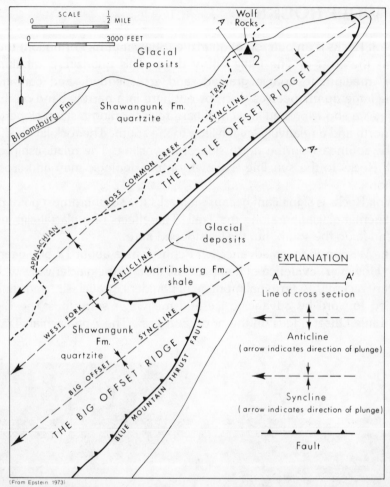

(From Epstein 1973)

Geologic map of Wolf Rocks, showing the different rock units folded into anticlines and synclines, surrounded by glacial deposits.

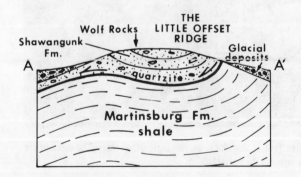

A geologic cross section through The Little Offset ridge and Wolf Rocks.

to the north of Wolf Rocks toward Fox Gap is partly on ground moraine made up of gravel, sand, and a few boulders. This mixture of rock types and sizes was left beneath retreating ice and is foreign to this area. In addition, at numerous places northeast of Fox Gap, the rock surface of Kittatinny Mountain has been scratched by glacial ice, producing glacial striations. As mentioned in the preceding section on the Delaware Water Gap, these lines on the bedrock surface help to show direction of ice movement. Most of these glacial scratches have weathered away, but some may still be found.

From here southward on the Trail, the effects that the glaciers had upon the climate and subsequent rock weathering may be seen, but the effects of direct contact with moving ice are seen only to the north.

3. WIND GAP

In *Northampton County, Pennsylvania*, written in 1939, Benjamin Miller noted (p. 152):

> Wind gaps are gaps or notches in mountains, cut by streams which were later diverted to other places. They are common features in the Appalachians and elsewhere where adjustments of stream courses to geologic structures have taken place. The name suggests some connection with air currents and at times their origin has been erroneously attributed to the wind. They have received their name because occasionally surface winds are turned or directed by these gaps to such an extent that they are noted by the local residents.

From the Delaware River to the Lehigh River, Totts Gap, Fox Gap, Wind Gap, Smith Gap, and Little Gap are all wind gaps. The position of each one is controlled by one or more geologic factors which result in weaknesses in the rocks that form the ridge. At Wind Gap, this controlling geologic factor is a zone of vertical, closely spaced fractures in the rock. Thus the quartzite at Wind Gap was more easily eroded than the quartzite in nearby areas.

Photograph courtesy of Malcolm L. White

View looking south across snow-covered hills of red shale to Wind Gap in Blue Mountain.

The locale has a long history of quarrying and mining of mineral resources. For a hundred years prior to the 1930's, iron mining was an active industry in the Lehigh Valley south of Blue Mountain. South of Allentown, in the Saucon Valley, one of the largest zinc deposits in the United States is mined at Friedensville. The most valuable mineral deposit in the vicinity is a pure limestone used in the manufacture of cement. An area from Egypt (Lehigh County) to Nazareth (Northampton County) in the Lehigh Valley is known as the "cement belt" of Pennsylvania.

Close to the Appalachian Trail, less than 2 miles southeast of Wind Gap, is the center of the second-largest slate-producing region in the United States. This area has a history of over 100 years of production. Today, among the remaining operating quarries, there are great numbers of deep, abandoned, water-filled slate quarries and mountainous piles of waste rock which dot the landscape of Pen Argyl, Bangor, Wind Gap, and many other towns and villages southeast of the Trail.

Slate, from the Martinsburg Formation (Ordovician age) occurring south of Wind Gap, is a fine-grained rock that can be split along flat cleavage planes into continuous slabs. These slabs are chipped or cut into various sizes and used in roofing, blackboards, school slates, electrical panels, billiard and laboratory table tops, bathroom fixtures, stair risers, and grave vaults. Some of it is crushed for insulation, paint filler, and slate roof granules. The industry has declined since the 1950's due to the widespread introduction of manufactured products and less expensive paper products, but modern output from the remaining quarries still accounts for significant production of dimension stone.

4. SMITH GAP

Smith Gap, the smallest wind gap along the Trail, is located on the mountain between the Delaware and Lehigh Rivers. It is hardly a gap at all and one may wonder why it is included with the more notable Wind Gap and Little Gap, among others.

However, Smith Gap is significant because of its relationship to other topographic features and what it suggests about the geologic history of this area. The topographic map shows the following: (1) Smith Gap is a small notch in a long, uniform ridge which exhibits little change for miles from Wind Gap to Little Gap; (2) the elevation of the floor of the gap is 1,550 feet, whereas the ridge to the south-west is only 60 feet higher (elevation 1,610); (3) the V- or U-shaped contour lines on the flanks of Blue Mountain north and south of the gap indicate old stream-cut channels leading from the gap to the valleys below; and (4) Aquashicola Creek in the narrow valley to the north flows westward to the Lehigh River, whereas Smith Gap lies between the Lehigh and Delaware River systems.

SMITH GAP

Photograph courtesy of Malcolm L. White

View looking north to the small depression of Smith Gap in Blue Mountain. Geologic conditions were not favorable for the development of a deeper gap here. Shale of the Martinsburg Formation underlies the valley floor in the foreground.

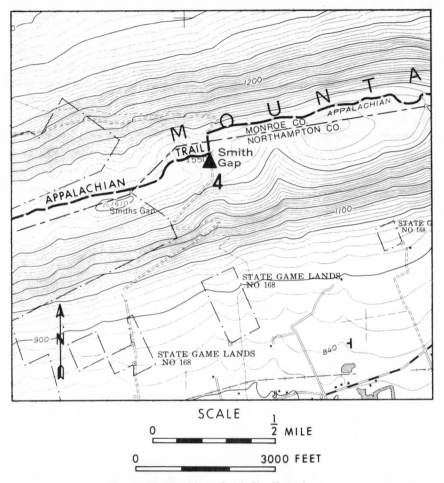

Topographic map of Smith Gap, in Blue Mountain.

The evidence shows that Smith Gap is high, near the ridge crest, and not well developed. Erosion by stream water on the long, linear bands of bedrock of varying resistance has been going on for a long period of geologic time. Evidence available now indicates that the pattern, which included water gaps and wind gaps, was profoundly affected by Pleistocene glaciations. Plate 1 shows that the earlier Illinoian glaciation occupied the valleys north and south of Smith Gap and that the more recent Wisconsinan glaciations extended through the valleys to the vicinity of Wind Gap. Also there is evidence for extensive glacial-lake development associated with the more recent glaciations.

Because Smith Gap is high and not well developed it was probably abandoned early by the stream that flowed through it. Prior to glaciation, the stream that "captured" the gap may have been in the Delaware River system, as appears to be the case with nearby Wind Gap. Subsequently, glacial activity blocked the flow to the northeast, and the relatively small Aquashicola Creek was diverted to the Lehigh River.

Finally, none of the structural geologic criteria for the location of gaps discussed in the sections on the Delaware Water Gap (site 1) and Wind Gap (site 3) are found at Smith Gap, so it may have been a bad place for a good gap to form. Geologic conditions for structural control of a gap were more favorable at Little Gap.

5. LITTLE GAP AND DEVILS POTATO PATCH

The Trail at this point affords the hiker still another look at a wind gap. Little Gap may be a misnomer, however, because it is not so little—there is a difference of more than 300 feet in elevation between the gap floor and the adjacent ridge crests. It is an excellent example of a wind gap and its origin is probably similar to that of Wind Gap and Smith Gap on the Trail.

Within the gap and just south of the Trail is one of the most outstanding geologic features in this area, Devils Potato Patch. The "Patch" is a boulder field that has accumulated as a result of the intense weathering (mostly freezing and thawing) of the rocks on the nearby ridge.

The boulder field is relatively flat and appears to be more than 10 feet thick. The field is made up of a jumbled assortment of loosely

View looking across Little Gap from the east. The rock is quartzite containing numerous quartz veins.

Intense weathering
(breakup) of rocks
on Blue Mountain

(Modified from Geyer, 1969)

packed boulders that have no fine material between them. The boulders themselves are yellow-gray quartzite and conglomerate (Shawangunk Formation).

Under near-ice (periglacial) conditions, the ground was permanently frozen, and freezing and thawing were severe on the surface during the summer months. The ridge to the east of the boulder field was

Devils Potato Patch. Boulders of quartzite and conglomerate form a level-surfaced boulder field in Little Gap. This is the northernmost of many outstanding periglacial features along the Appalachian Trail in Pennsylvania.

subjected to this intense weathering and was slowly and repeatedly broken into blocklike masses. Gradually they moved downslope, grinding against each other, and some of their angular corners were worn away to produce the flat boulder field that we see today.

When the glacier vanished and the climate warmed, all movement stopped. Fine sand and clay between the boulders have since been washed away, leaving the open spaces you see. Devils Potato Patch marks the northernmost of many periglacial features along the Trail from here southward.

Throughout the ridges surrounding the gap and on the west side of the gap, occasional outcrops of quartzite and conglomerate may be found. But, for the most part, the sides of Blue Mountain are covered with blocky rock talus weathered from the ridge crest in much the same way and during the same time in which the boulder field was formed.

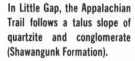

In Little Gap, the Appalachian Trail follows a talus slope of quartzite and conglomerate (Shawangunk Formation).

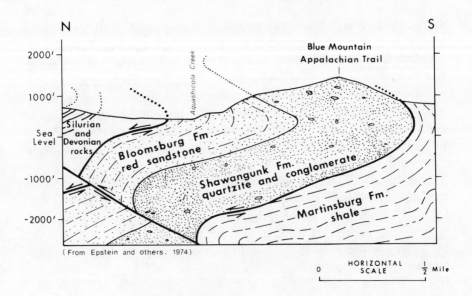

(From Epstein and others. 1974)

HORIZONTAL
SCALE

0   ½ Mile

Geologic cross section through Blue Mountain at Little Gap.

During past geologic time, Blue Mountain has been subjected to severe mountain-building forces, especially in this area. The hard,
strong quartzite layers have been turned upside down through intense
folding and faulting. Another result of all of this geologic activity is the
breakup of the rock itself. It is this fracturing of the quartzite that made
it an easy target, during the periglacial climatic conditions, for further
breakdown to the rock features we see.

6. LEHIGH GAP AND DEVILS PULPIT

Lehigh Gap is another spectacular water gap. However, in addition to the gap, two unusual features here may be of interest to the hiker. First, there is a lack of mature vegetation on hillslopes of this gap, and second, there is an outstanding rock pinnacle on the west side called the Devils Pulpit.

The northeast side of Lehigh Gap, almost barren of vegetation.

To explain the absence of vegetation we must start several miles south of the Trail near Bethlehem, where there is a valuable zinc deposit being mined at Friedensville. For many years at Palmerton, just north of the gap, ore from this mine has been processed in furnaces for the production of a zinc concentrate. It is believed by some that flue gases from this operation have blown through the gap for a long time, killing most of the trees. The amount of effluent has decreased in recent years and resistant vegetation has been planted.

Early settlers named many apparently inaccessible, rough and rocky places after the Devil. Along the Appalachian Trail on Blue Mountain we find the Devils Potato Patch, the Devils Pulpit, and the

The Devils Pulpit, called "Die Teufels Kanzel" by early German settlers, is an erosion remnant of quartzite standing as a podium on the west side of the water gap. Notice also the barren landscape.

Devils Race Course. Here, the Devils Pulpit is an erosion remnant of gently dipping quartzite in the form of a giant pulpit in the sky, called "Die Teufels Kanzel" by early German settlers.

7. LEHIGH FURNACE GAP

Lehigh Furnace Gap is another picturesque wind gap in Blue Mountain. The name of this gap is taken from the Lehigh Furnace, a long-abandoned iron furnace at the foot of Blue Mountain near here. Built in 1826, it was operated until 1870 utilizing iron ore and limestone from near Allentown. Charcoal to fire the furnace was made locally from mountain timber. Iron mining started in Lehigh County in 1809, and the greatest activity was from the end of the Civil War to about 1885. Shortly after the turn of the century, mining and processing of ore ceased because of the discovery of large iron deposits in the Lake Superior district.

Ocher (a mixture of limonite [a weathered iron mineral] and clay) and umber (the same material as ocher, with the addition of manganese oxide) have been extensively mined in the valley north of Blue Mountain. They were used as a filler and in the manufacture of paint. Farther to the east, iron carbonate associated with limestone was mined as paint ore and used in the manufacture of paint products.

Remains of the Lehigh Furnace at the foot of Blue Mountain.

The conglomerate at Lehigh Furnace Gap is intensely fractured and broken. A rocky footpath results.

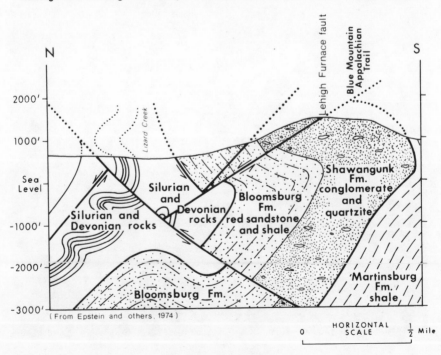

(From Epstein and others, 1974)

Geologic cross section through Blue Mountain at Lehigh Furnace Gap.

The geology of the Appalachian Trail in this vicinity is complex, characterized by several rock formations that have been broken by numerous faults and deformed by folding. Folding and faulting are more intense along this portion of Blue Mountain than to the northeast of Lehigh Gap. As a result, the rocks are intensely fractured and broken, forming a rocky footpath. A geologic cross section through Blue Mountain at the gap graphically portrays the complex geology. Major segments of the mountain have moved along fault zones, and the rock that forms these segments is folded and twisted so that much of it is upside down.

8. BAKE OVEN KNOB

Because of the outstanding view to the south from its summit and perhaps because of its catchy name, Bake Oven Knob has become a famous place along the Appalachian Trail. In an early publication describing topographic names, the knob is listed under the subheading "Names for fancied resemblances" (Miller and others, 1941). It is said that a depression in the flank of Blue Mountain below this location resembles an outdoor bake oven, and hence the name Bake Oven Knob for the rocky peak above it.

The rocks here are gray conglomerate and quartzite, containing pebbles as much as 6 inches long. The rocks are resistant to erosion, brittle, and extensively broken into large blocks.

The geologic map shows that the outcrop pattern curves south in response to movement along the Lehigh Furnace Gap fault. The cross section shows the vertical movement on this fault and the relationship of the fault to overturned rock units that make up Blue Mountain. The horizontal movement along the fault (see geologic map, Plate 1) indicates a considerable shift of the northern block to the northeast. The

Bake Oven Knob from the southeast is exposed as a rampart on the crest of Blue Mountain. The Bake Oven is below it in the trees.

combination of folded and erosion-resistant rocks, faulting, and, finally, bending of the entire mountain range resulted in this broken knob.

A cross section through Blue Mountain across the Appalachian Trail about one-half mile east of Bake Oven Knob shows beds overturned in steep folds and major faults, which together contributed to the formation of the knob.

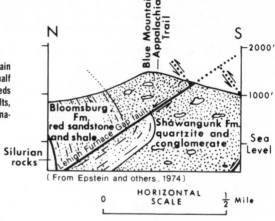

From a distance Bake Oven Knob is deceptively gentle in appearance. Not until one walks across its craggy spine is its rocky, angular nature seen by the hiker.

9. BEARS ROCKS

Mistakenly shown as "Baers Rocks" on topographic maps, Bears Rocks, outcrops of quartzite on the crest of Blue Mountain, is said to have housed many bears in the past. The rock is gray, medium-bedded quartzite that has well-displayed bedding and crossbedding.

This site is an excellent example of weathering characteristics of rock all along the crest of Blue (Kittatinny) Mountain. As the mountains slowly formed through folding and faulting approximately 250 million years ago, they started to weather and erode. Mountain building and weathering were active at the same time, but building forces were stronger than weathering, so that a range of mountains was constructed. Shale and limestone underlie the valleys and were eroded more quickly than the hard quartzite and conglomerate of the ridges. As mountain building slowed and weathering continued, the resulting difference in elevation left serrated ridges with upturned edges of hard, fractured rock. Water from precipitation froze and thawed in the fractures, slowly breaking off block after block of rock. The spacing of bedding planes is important to this process because fracturing along these planes occurs during weathering and, together with joint spacing, determines the average size of pieces that break away from the outcrop. As they break away, these pieces build up on one another and creep downslope by gravity, covering the hillsides with angular boulders.

Bears Rocks on Blue Mountain. Take a few minutes to study the bedding (horizontal layers) and the jointing (vertical fractures). Note the crossbedding in this exposure.

10. THE CLIFFS

The Cliffs are steeply dipping layers of quartzite and conglomerate that appear like a handmade stone wall on the crest of Blue Mountain. The Appalachian Trail follows this "stone wall" for about one-half mile. The prominence of this exposure serves as a good example of the persistence and outstanding nature of the quartzite and conglomerate ridge from the Delaware River to the Susquehanna River.

The Cliffs form a "stone wall" along the narrow crest of Blue Mountain. The rock is quartzite and conglomerate of the Shawangunk Formation, overturned and dipping steeply to the south (right).

Buried beneath the talus on the steep hillside to the south of The Cliffs is the contact between the shale and sandstone (Martinsburg Formation) in the valley and the quartzite (Shawangunk Formation) on the hilltop. The contact is a zone of detachment between two major bedrock units of differing character along which there has been regional movement. It is a widespread fault zone, having been recognized at this contact, where exposed, over most of its length in Pennsylvania. The folded thrust fault that occurs in this zone is called the Blue Mountain fault and is shown in the accompanying cross section.

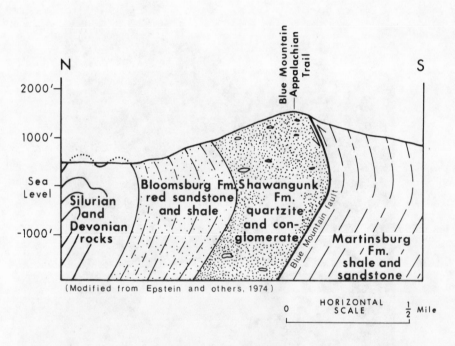

Geologic cross section at The Cliffs.

11. DANS PULPIT

Dans Pulpit stands on the crest of Blue Mountain. The origin of
Dans Pulpit is related to a fault that extends northeast-southwest, having offset the quartzite of Blue Mountain. The fault may actually be
longer than shown on the geologic map, but any evidence of its extension is buried under deposits of boulders on the mountain flanks.

Overlooks to the south are common along the Trail here. At Dans
Pulpit the view to the south is across a northern extension of the Great
Valley to another ridge of quartzite. This narrow valley segment has
been eroded through folded and deformed shale and sandstone of the
Martinsburg Formation. These more easily deformed, relatively weak
rocks have been compressed between the vise-like jaws of the hard

Dans Pulpit is a promontory
of conglomeratic quartzite
(Shawangunk Formation). This
rock is about 430 million years
old.

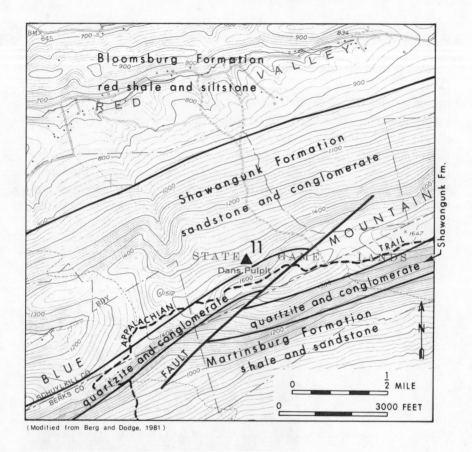

(Modified from Berg and Dodge, 1981)

The geologic map of the Dans Pulpit area is characterized by a short fault that cuts through Blue Mountain at Dans Pulpit, offsetting the quartzite ridge.

quartzite (Shawangunk Formation) ridges on either side during periods of folding many hundreds of millions of years ago.

Between Dans Pulpit and The Cliffs on the Appalachian Trail, a tricounty marker stands on the broken ridge. Here, too, faulting has caused a portion of the crest to be extensively broken.

The valley south of Dans Pulpit is underlain by squeezed and faulted shale and sandstone (Martinsburg Formation). Quartzite of the Shawangunk Formation underlies the ridges.

The tri-county marker designating the corners of Lehigh, Berks, and Schuylkill Counties is on a ridge of broken quartzite between The Cliffs and Dans Pulpit.

12. RIVER OF ROCKS AND HAWK MOUNTAIN

The River of Rocks is a long, narrow boulder field, an accumulation of boulders exposed at the head of a valley between ridges of quartzite that surround it like an amphitheater. The "River" is part of a larger field which is mostly covered with vegetation. The boulders are now stable, and there is little evidence of recent movement. The field is thought to have formed in a periglacial climate when intense freeze-thaw weathering acted on strongly jointed quartzite cliffs above, loosening large and small fragments and allowing them to move downhill by creep and solifluction. Solifluction is slow, viscous downslope movement of saturated soil and unsorted rock debris. Most of the boulder field, extending from ridge to ridge, is unsorted; the large boulders and smaller fragments are interlocked in a rocky soil that supports vegetation. However, stream action at the bottom of the trough

River of Rocks.

Photograph courtesy of William H. Bolles

of deposition, marked by the River of Rocks, has removed finer fragments, leaving an accumulation of large boulders that have voids between them and a stream underneath, a local environment that does not support vegetation.

The Hawk Mountain Lookout is a spectacular location near the Trail. The old saying applies here: On a clear day you can see forever! This lookout is well known for being one of the nation's outstanding vantage points along the eastern migration flyway for hawks and eagles.

The scenery is typical of the ridges and valleys of central Pennsylvania. Pause a moment and reflect on the diversity of the scenery along the Trail. Then remember that what you see is the result of the character of the rocks beneath.

Take a few more moments and look closely at the rock you are sitting on. Look for individual quartz "sand" grains "cemented" together with quartz. This rock was once a sandstone, but intense heat and pressure during burial under other sediments have changed it to a metamorphic rock called quartzite, which is seen along the Trail for many miles. The major difference between the two rock types, sandstone and quartzite, may be seen here. Quartzite breaks across the individual grains, whereas sandstone breaks around each grain.

View looking northwest from Hawk Mountain Lookout across a valley underlain by red shale of the Bloomsburg Formation to a ridge of sandstone (Catskill Formation) in the distance.

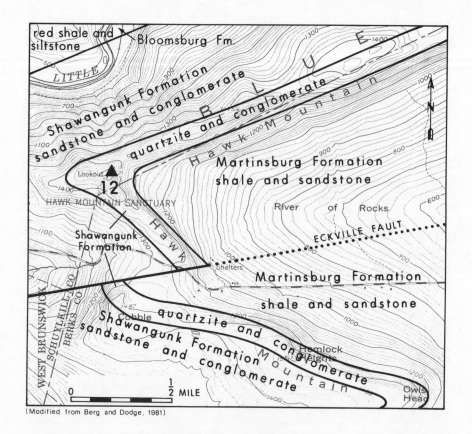

(Modified from Berg and Dodge, 1981)

Geologic map of the Hawk Mountain area.

If you were in an airplane flying over this location you would see evidence of a much larger geologic feature—a fault that broke and separated Hawk Mountain. The geologic map of the area shows the location of this fault.

Quarrying for sand was done near the crest of Hawk Mountain west of the lookout. Perhaps as a result of intense folding and faulting of the rocks, the quartzite here is friable (crumbly); part of it is disaggregating and breaks away easily from the outcrop. Between 1870 and 1900, a gravity tramway was used to transport sand and silica rock from this source to a railroad below on the Little Schuylkill River.

13. THE PINNACLE

The Pinnacle is a bare-rock projection of a highly fractured quartzite layer, almost flat lying, at the nose of a large syncline. This synclinal fold has moved over the underlying deformed shale and sandstone along a large thrust fault. A geologic cross section through the moun-

The rocks of the valley south of The Pinnacle are intensely folded and faulted shale and sandstone. Spitzenberg Hill is in the background to the left.

tains and valleys from Dans Pulpit on the north to Blue Rocks on the south, in the vicinity of The Pinnacle, shows the geologic relationships within this complicated structure.

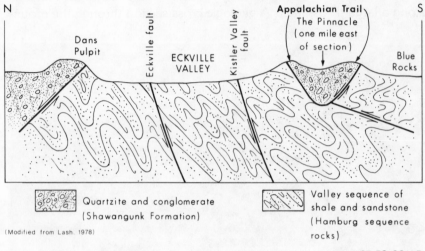

(Modified from Lash, 1978)

Quartzite and conglomerate (Shawangunk Formation)

Valley sequence of shale and sandstone (Hamburg sequence rocks)

NOT TO SCALE

Geologic cross section from Dans Pulpit across the Eckville Valley to The Pinnacle.

East of The Pinnacle a cone-shaped hill called Spitzenberg stands alone in the valley. The origin and age of the rocks in this hill have been argued by geologists for 100 years. It consists of sandstone and conglomerate, which is mapped on Plate 1 as the Ordovician Juniata and Bald Eagle Formations, undifferentiated. However, other interpretations indicate that it is part of the Martinsburg Formation.

A cave and narrow passages under The Pinnacle near its crest are joint-block openings where bedrock has been fractured and moved apart during periods of great rock folding and subsequent erosion.

Fossils of animal and plant parts have not been found in the quartzite of the Shawangunk Formation or in the Tuscarora Formation. There are, however, trace fossils, which are tracks, trails, or burrows left behind as evidence of past animals that had no hard skeletal parts capable of being fossilized. Branching tubes and small ridges on bedding-plane surfaces at The Pinnacle are *Arthrophycus*, which are thought to be worm tubes, evidence perhaps of worms inhabiting nearshore environments during the Silurian Period.

Ledges of quartzite form The Pinnacle. Caves beneath The Pinnacle are "joint block" openings.

ARTHROPHYCUS
HORIZONTAL WORM BURROWS

EARLY SILURIAN AGE
(420 million years B.P.)

TUSCARORA FORMATION

The only fossils that have been found in the bedrock along Blue and Kittatinny Mountains are trace fossils which appear to be ancient worm tubes. They are well exposed in the gently dipping beds at The Pinnacle.

14. PULPIT ROCK AND BLUE ROCKS

Pulpit Rock is an outstanding platform of quartzite jutting from the crest of Blue Mountain. Study the rock surfaces closely at this site. Some exhibit polished, grooved surfaces, or slickensides, formed when one block moved against another or when a fault caused rock surfaces to slide against each other during mountain building. Other rock faces exhibit a glittering array of small quartz crystals that look like a sugar frosting. These crystals formed some 280 million years ago from silica-rich water which moved through tight, narrow fractures in the rock when it was deeply buried beneath thick layers of other rock. These two features, slickensides and quartz crystals, are also well displayed between this site and The Pinnacle (site 13).

To the east of Pulpit Rock and the Trail lies the Blue Rocks boulder field. This boulder field is one of the largest and most spectacular in Pennsylvania. The surrounding quartzite ridge from Pulpit Rock to The Pinnacle is the source area for an extensive boulder rubble deposit, of which the Blue Rocks boulder field is the center.

It is hard to imagine blocks of rock and boulders that cover ridges and valleys moving very much today; in a human lifetime none of them do move very much, even on steep hillsides. They are mostly interlocked, like pieces of a large, completed rock puzzle spread across ridges and adjacent mountain valleys. However, when the margin of the continental glacier was nearby, a climate existed in which ice wedging along narrow fracture openings was intense, effectively quarrying blocks from the ridges in an environment in which the fresh pieces could move downhill by creep and solifluction. The sequence of photographs shows what occurred. (A) The quartzite on the ridge opened along a joint, as shown where one section of rock moved far enough from the other to allow the Trail to pass through. (B) Blocks broken from the ridge slid downhill, forming the interlocking mass. (C) Blocks that started their journey from the ridge crest reached the valley bottom, where stream action washed away finer material and left an accumulation of large interlocking blocks, exemplified by the Blue Rocks boulder field.

On the Trail, at the foot of Blue Mountain southwest of Pulpit Rock, the remains of Windsor Furnace show only as a small ore pit, building foundations, and scattered furnace slag. Some iron ore may have occurred here, but the primary local resources were timber for charcoal and fast-flowing streams to drive a water wheel for the blast furnace. As at Lehigh Furnace, the iron ore probably came from the limestone valley near Allentown.

(A) The Appalachian Trail passes through a crack (joint) in the rocks. Intense freeze-thaw weathering during the Great Ice Age opened a passageway along this joint in the quartzite.

(B) Pulpit Rock and boulder talus weathered from the cliff.

(C) The Blue Rocks boulder field.

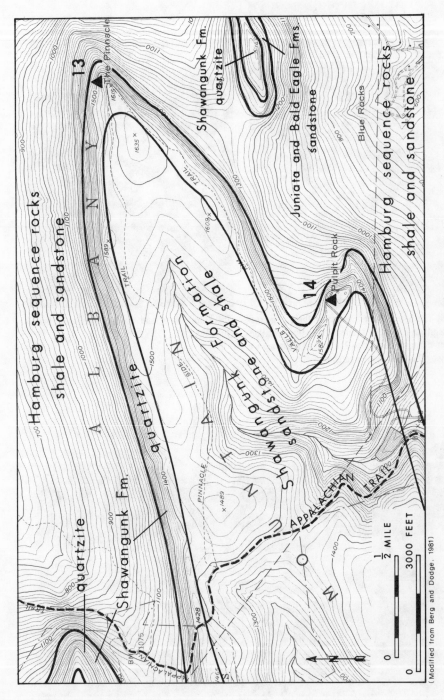

Geologic map of the vicinity of Pulpit Rock and The Pinnacle.

(Modified from Berg and Dodge, 1981)

15. AUBURN LOOKOUT

This lookout provides a beautiful panoramic view northward from Blue Mountain. Views in this direction are uncommon between Hawk Mountain and Swatara Gap. Rock weathering of the ridges is well exemplified at this location, as shown in the photograph. Boulders man-

The view north from Auburn Lookout.

An extensive deposit of quartzite boulders covers the slope below Auburn Lookout.

tle the slopes, forming an extensive deposit of large, angular fragments. The natural slope is stable and not subject to landsliding, because the rough-hewn pieces become interlocked.

The Trail has been following a major fold structure, a syncline, since it crossed The Pinnacle. However, here the fold dies out and faulting becomes more pronounced along the Trail between Port Clinton and the Monroe Valley Overlook. The faults generally are parallel to the mountain trend and have caused bedrock units to be repeated, widening the ridge. One of these faults is nicely exposed in outcrop close to the Appalachian Trail along an abandoned railroad grade on the west side of the Schuylkill River, 2,000 feet south of the junction of the Trail and the railroad grade. Slickensides, which are polished and smoothly striated rock surfaces caused by friction along a fault, are beautifully displayed here in one of the best exposures of them in the Appalachian Mountains of Pennsylvania.

16. SCHUBERTS GAP

The Appalachian Trail descends into and climbs out of Schuberts Gap following a course similar to that in many other gaps. There is a difference here, however, in that the gap appears as only half of a water gap—on the north side, where the other part of the gap should be, there is an extra ridge.

A long fault, shown on the geologic map, has caused Blue Mountain to repeat itself for a distance of about 4-1/2 miles in this area. The attitude and movement of the fault and quartzite ridges are shown in the simplified block diagram on the following page. The quartzite formation on the north has moved up with respect to that on the south. This fault splits into two branches, becoming part of the fault system that formed Round Head (site 17) to the west.

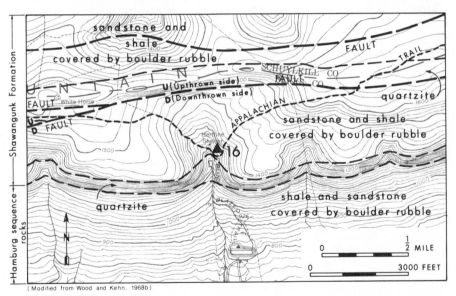

(Modified from Wood and Kehn, 1968b)

The geologic map of the Schuberts Gap area shows the quartzite ridge repeated by faulting. The stream flowing south from Schuberts Gap has cut part of the way through one ridge.

A stream that has a steep gradient, such as the one flowing south from Schuberts Gap, cuts upstream, lengthening its valley by headward erosion. This headward erosion is caused by groundwater sapping and by slumping of blocky rock material into the valley. In the future, the stream might cut through Blue Mountain by headward erosion to form a complete water gap.

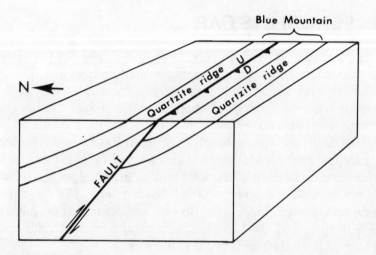

Repeating of a ridge as a result of movement along a fault is shown in this schematic block diagram.

17. ROUND HEAD

Round Head overlook is one-half mile south of the Appalachian Trail on a blue-blazed side trail. The geology of this feature and surrounding area is interesting and well worth the short side trip.

Accumulations of boulder rubble make a rocky footpath in the vicinity of Round Head. Weathering of upturned, bare outcrops of quartzite and conglomerate along Blue Mountain produces many boulders. Where the quartzite formation is further folded and faulted, as it is here, the quantity of boulder rubble is increased. Because of this, the Appalachian Trail here is about as rocky as it can get.

Two faults and an associated fold have broken and offset the mountain, as shown on the geologic map. Round Head is a small syncline, the axis of which trends northeast-southwest. It is separated from Blue Mountain by a fault along which Blue Mountain went up and Round Head down. A second fault, north of the first, displaced Blue Mountain and is the same one that affects the mountain at Schuberts Gap.

Exact positions of the faults and contacts between bedrock units are interpreted by the field geologist. There is so much boulder rubble that exact geologic relations are masked and it becomes necessary to rely on experience and the examination of rock units that are not covered in adjacent areas in order to correctly interpret the geology here.

View looking west from Round Head toward Swatara Gap.

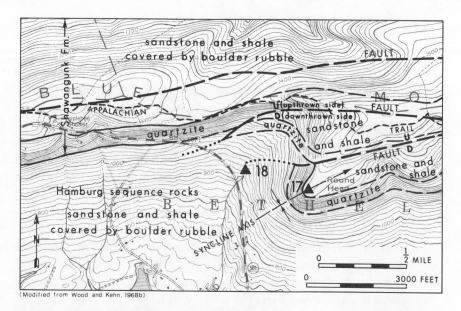

(Modified from Wood and Kehn, 1968b)

Geologic map of the vicinity of Round Head and The Kessel. Faults have broken the ridge in this area, resulting in a significant offset along the crest.

Nested rocks form a boulder rubble deposit on the nose of Round Head.

18. THE KESSEL

"Die Kessel" is early German for "The Kettle," a bowl-shaped land-
form surrounded by Round Head and the south flank of Blue Mountain. As at Round Head (site 17), the weathering of the quartzite for-

Large blocks of quartzite cover the sides of The Kessel with boulder rubble.

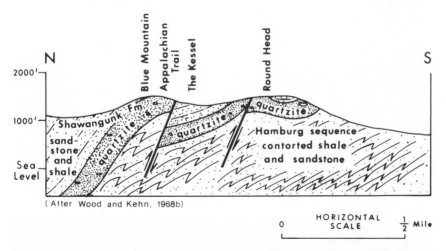

A geologic cross section through the bedrock that forms Blue Mountain, The Kessel, and Round Head
shows two faults, one of which is on the north side of The Kessel parallel to the Appalachian Trail.

mation on the ridge crest has produced large masses of boulder rubble on the slopes of Blue Mountain.

Two faults join into one east of Round Head, extending past Schuberts Gap almost to Port Clinton. Broadening of the ridge crest associated with these faults begins to diminish southward from The Kessel, and the ridge narrows significantly beyond Monroe Valley Overlook.

19. MONROE VALLEY OVERLOOK

Geologic relations at this location are best displayed on the full-color geologic map (Plate 1). A small offset in the quartzite and sandstone ridge in the vicinity of the Trail is mapped as a tight anticlinal fold. A secondary ridge formed by this fold can be seen from the top of Blue Mountain, looking north. Field evidence and the study of air photographs indicate that there is also a fault on the north side of the mountain, the trace of which is northeast, following a tight, narrow valley between the mountain and the secondary ridge. Using the interpretation of those faults mapped to the east, this one is most likely a reverse fault, the surface of which dips to the north.

South of Blue Mountain a ridge called Little Mountain extends for about 5-1/2 miles from the Berks-Lebanon County line to Swatara Gap. It terminates abruptly on each end and, except for a small gap through it, is a continuous linear feature. Rather than being the result of folding or faulting, as so many features along the Trail are, this apparently anomalous ridge is the result of changes in the sedimentary

View looking south from Monroe Valley Overlook. Little Mountain is in the background.

material that became the Martinsburg Formation. During deposition of the Ordovician shale sequence, which now makes up this part of the Great Valley, several zones of poorly sorted, coarse-grained sandstone were included in the sediment. These were hardened into well-indurated rock, which, through deformation and erosion, were left as ridges above the surrounding shale terrain. Little Mountain is one of these ridges visible from the Trail.

The geologic map (Plate 1) shows some lines of arbitrary cutoff between geologic formations in this area, indicating changes in name from one formation to the other. These represent points at which local detailed mapping has been extended as far as is reasonable, and in the case of the Shawangunk-Tuscarora cutoff the boundary represents a termination of nomenclature carried from the east and replacement by nomenclature carried from the west.

Southwest from this location the Appalachian Trail follows the quartzite ridge of Blue Mountain to Swatara Gap. The Trail then leaves the ridge, circling to the north, and Blue Mountain is not encountered again until south of Cove Mountain (site 29). The quartzite ridge, however, continues in a narrow arc across southeastern Pennsylvania, extending through Maryland and West Virginia into Virginia. The Tuscarora and Big Blue Trails follow this ridge.

20. SWATARA GAP

Swatara Gap is an outstanding water gap in Blue Mountain. As stated before, the actual carving of the gap is a dynamic process. Even today, the abrasive action of hard, angular rock fragments carried by Swatara Creek is wearing away the streambed. The resistant quartzite and sandstone layers are thinner here and weakened by a series of fractures. This localized weakness permitted faster erosion and the forming of the gap.

At Swatara Gap, the importance of water gaps to the development of transportation in Pennsylvania is well exemplified. An Indian path, canal, aqueduct, railroad, local road, interstate highway, and now a national scenic trail have all passed through the gap.

For the first time since entering the state at the Delaware Water Gap, the Trail leaves the quartzite (Tuscarora Formation and Shawangunk Formation) ridge and goes northwest across other rock types. The rocky trail has been following the crest of the first mountain (Blue Mountain and Kittatinny Mountain) of the Appalachian Mountain section of the Valley and Ridge province from New Jersey to this water gap. It now leaves that crest, crossing steeply dipping, folded

View looking north at Swatara Gap in Blue Mountain.

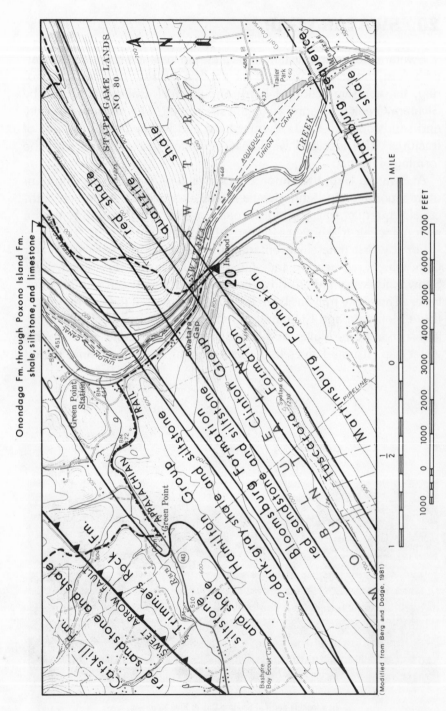

Geologic map of the Swatara Gap area. Swatara Creek has cut across steeply dipping, nearly vertical rocks to form the gap.

Steeply dipping quartzite and shale crop out at Swatara Gap along Interstate 81 southbound.

and faulted rock units of shale, siltstone, and sandstone through a rolling valley north of Blue Mountain.

An exposure of shale (Martinsburg Formation, Ordovician age) along the west side of the Trail on Pa. Route 72, under the Interstate 81 bridge, is a famous fossil-collecting locality. This site contains the largest abundance of the Ordovician trilobite *Cryptolithus* in Pennsylvania. Equally important is the presence of the rare Ordovician starfish *Hallaster*.

Cryptolithus   *Hallaster*

Most of the fossils found here are molds of the original animal. Complete specimens are unusual, but fragments are numerous. A bright orange stain on the fossils is common, and is a helpful feature to spot when hunting for them.

The abundance of fossils here indicates that the inland sea where these muds were first deposited was rather shallow, that the water was fairly clear, and that there was sufficient food to support this prolific animal life.

21. RAUSCH GAP

Rausch Gap is a water gap in Sharp Mountain that has an interesting history relating to geology. This mountain marks the southern edge of the important Pennsylvania anthracite region, a hard-coal district that extends for another 100 miles to the northeast. Anthracite has been mined in this area since 1769, and the greatest activity at Rausch Gap started in 1826. The Schuylkill and Susquehanna Railroad Company and the Dauphin and Susquehanna Coal Company operated locally from 1850 to 1870. But on this extreme edge of the otherwise rich coal basin, mining was marginal and the coal was not of great value.

The intersection of the Trail and the railroad grade at Rausch Gap was the center of a community by the same name that started as a lumbering and coal-mining center in the 1840's and was abandoned by the early 1870's. The village had a maximum population of about 1,000 people. The railroad, started in 1852 and put into operation in 1854, was used to transport coal, lumber, freight, and passengers between Pottsville and Dauphin. It ceased operating in 1933. In 1946 tracks and ties were removed, leaving, in this area, a 17.5-mile

This hand-built wall of conglomerate (Pottsville Group) is the remains of a railroad bridge abutment. The railroad once served the mines of Rausch Gap.

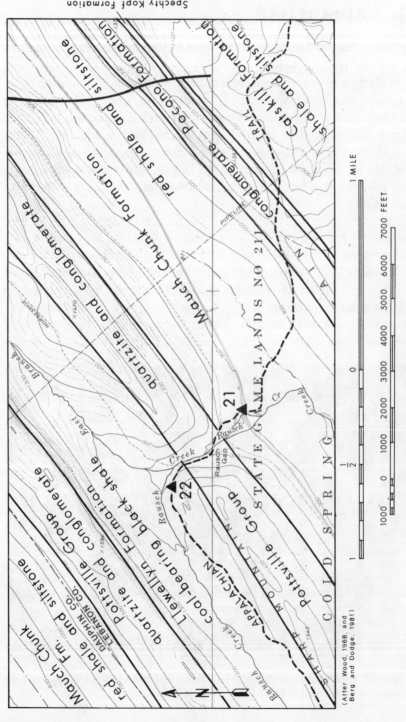

Geologic map of the Appalachian Trail in the vicinity of Rausch Gap.

(After Wood, 1968, and
Berg and Dodge, 1981)

recreational path from Ellendale Forge on the west to Gold Mine Road on the east.

Between Swatara Gap and Rausch Gap the Trail traverses mountains underlain by hard layers of sandstone, conglomerate, and quartzite, which are younger than those of Blue Mountain to the south (Plate 1). These parallel ridges are more resistant to weathering than the shale and siltstone of the adjacent valleys. At Rausch Gap the Trail crosses the hard, resistant quartzite and conglomerate beds of the Pottsville Group in Sharp Mountain, then descends into Rausch Creek valley, underlain by black, coaly shale and veins of anthracite coal (Llewellyn Formation). These beds of anthracite occur in tight folds that are extremely broken, faulted, and deformed. Individual coal beds are discontinuous and often mixed with black shale. The poor

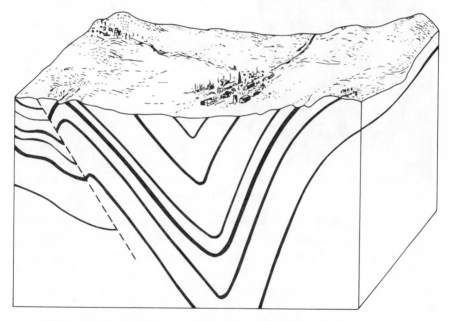

Schematic block diagram of the anthracite region showing folded and faulted coal beds.

quality of the coal at this locality, coupled with the thin and discontinuous nature of the coal veins, led to the short life of the mining town, Rausch Gap. By the early 1870's, more accessible and higher grade coal deposits were being developed to the northeast.

22. STRIP MINE

The only open-pit coal mine on the Appalachian Trail in Pennsylvania is along Sharp Mountain at the north end of Rausch Gap (see site 21 for exact location). In about 1935, the Mammoth coal seam was strip mined at this location. Strip mining by use of a drag line across the surface took place long after the underground mines had been abandoned.

Strip mine on the Mammoth coal seam.

The cross-section sketch shows the geologic positions of the Mammoth coal on the south limb of a syncline. In the south wall of the strip mine you can see the steeply south dipping rocks. Westward along the Trail this tightly folded syncline containing coal gets narrower and narrower and finally pinches out and disappears altogether as shown on Plate 1.

The long-abandoned underground mines in the vicinity of Rausch Gap are marked by walled or collapsed entries in the hillsides near the Trail. These seldom-seen mine entrances were almost forgotten until Hurricane Agnes occurred in 1972. With a dramatic rise in the groundwater level, suddenly new drainage paths from some of the mine entrances were created. Acid mine drainage flows from these mine portals into adjacent streams.

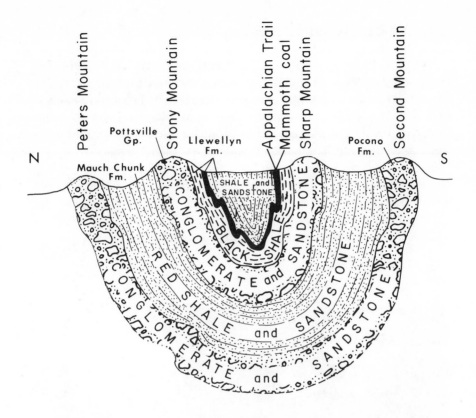

Cross section of the Mammoth coal seam and surrounding rocks in the south limb of a syncline.

Acid mine drainage is a significant environmental problem in Pennsylvania. It is the result of a chemical reaction between water, oxygen, and pyrite (FeS_2). Pyrite, also called fool's gold, is a common mineral disseminated through the coal and adjacent black shale. A two-stage chemical reaction causes the formation of sulfuric acid and the precipitation of iron as a yellow-orange coating. As the Trail leaves Rausch Gap, southbound, it crosses a small stream. The water is clear but acidic and the rocky bottom is coated with "yellow-boy."

23. YELLOW SPRINGS

Yellow Springs is the site of another ghost town. A once-thriving community engaged in coal mining was active here in the mid-1800's. One underground mine was worked from 1849 to 1853 and a second from 1850 to 1851. Coal from the mines was lowered down a long incline to a breaker at the railroad in Stony Creek Valley; from there it went by rail to Dauphin on the Susquehanna River.

Near one mine opening north of the Trail at Yellow Springs is a stone foundation beside a 25-foot-high stone tower. Originally the tower may have been as much as 35 feet high when in use in the 1850's. Although sometimes described as a mysterious obelisk in the woods, it was used for underground mine ventilation. A mining engi-

A stone tower stands as an eerie ruin in the woods north of the ghost town of Yellow Springs.

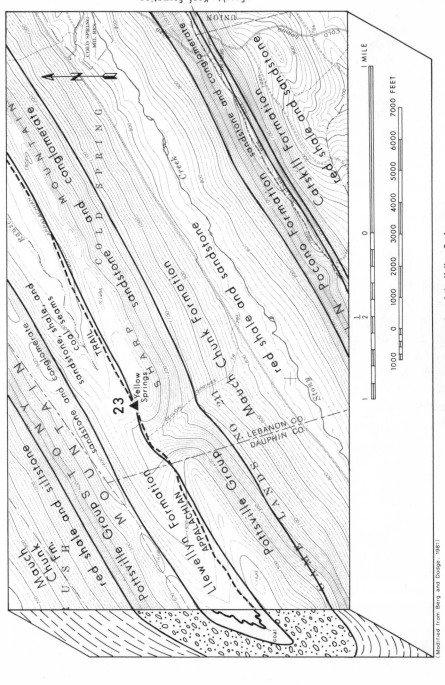

Geologic map and cross section of the Yellow Springs area.

(Modified from Berg and Dodge, 1981)

neers' handbook notes the following under mine ventilation (Peele, 1927, p. 1212):

> When the natural difference in density of downcast and upcast currents is insufficient to produce the required circulation, this difference may be increased by artificial heat. . . .To increase height of air column affected, a chimney may be built at top of upcast shaft.

The chimney at Yellow Springs was a stack for a steam engine and also was used for mine ventilation. There is a cast-iron pipe from the mine shaft to the base of the chimney. A steam-engine boiler fire augmented the draft and increased air flow from the mine, improving ventilation.

The geology here is characterized by a tight, pinched syncline, a narrowing extension of the one at Rausch Gap. A geologic cross section of the valley at Yellow Springs would be similar to the one at Rausch Gap except that the fold would be tightened. As the coal beds were folded and pinched out, they were badly broken, decreased in thickness, and lost continuity. The quality of the coal was poor. This was reflected in the short history of mining at Yellow Springs; also, the quantity of coal available was less. Where there were 12 coal seams at Rausch Gap, there are only five thin ones reported here.

Leaving Yellow Springs, the Trail (southbound) is on brown to gray sandstone and gray to black shale of the Llewellyn Formation. Shortly it crosses over the north flank of the syncline onto hard, gray sandstone and conglomerate layers of the Pottsville Group. However, almost none of the Llewellyn Formation is exposed because thick boulder rubble (colluvium) of conglomerate from Sharp and Stony Mountains has covered it in this narrow portion of the syncline.

24. DE HART RESERVOIR OVERLOOK AND DEVILS RACE COURSE

From Yellow Springs, the Trail crosses the crest of Stony Mountain and starts a descent into the valley of Clark Creek. The crest of Stony Mountain is underlain by conglomerate of the Pottsville Group, hard and resistant to erosion, whereas Clark Valley is underlain by red shale and siltstone beds of the Mauch Chunk Formation, which is less resistant and easily eroded. Just over the crest, an overlook provides a breathtaking view to the north toward the DeHart Reservoir and Dam on Clark Creek. The break in the trees at the Trail's edge marks a waste dump at the mouth of an abandoned coal mine on the southwest side of the Trail. This is the southernmost coal mine on the Appalachian Trail in Pennsylvania; it is in the upper portion of the Pottsville Group, which contains a few minable coals. Acid mine drainage is present here as at Rausch Gap.

The plant fossil *Lepidodendron*, representative of the Pennsylvanian geologic time period, can be found in the mine dump on the surface of shale slabs.

One of the most outstanding geologic features near the Trail is Devils Race Course. The very narrow, constricted valley south of the Trail, coupled with a gap in the south flank of the valley at Sharp Mountain, provides both an excellent source area of steeply dipping,

View looking northeast from the Stony Mountain lookout tower. Peters Mountain is in the background, and the DeHart Reservoir and Dam are in the center.

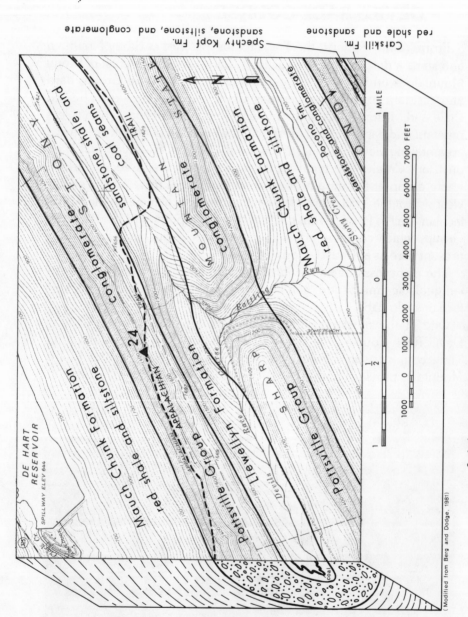

Geologic map and cross section of the DeHart Reservoir Overlook area.

(Modified from Berg and Dodge, 1981)

hard, fractured rock layers and a trough for accumulation of boulders extending down through the gap in Sharp Mountain.

Devils Race Course is a long, narrow boulder field (approximately 3,500 feet long and 120 feet wide) consisting of interlocking angular and subrounded blocks of conglomerate that do not have appreciable amounts of finer material between them. The stream of Rattling Run can be heard beneath the boulders, but is hidden from view.

The Devils Race Course, at the head of Rattling Run between Sharp and Stony Mountains, is an accumulation of boulders weathered from the adjacent ridges.

The surface of the field has a gradient ranging from 1-1/2 to 4-1/2 degrees. Although it is presently stable, there is evidence of past downslope movement shown in the "fabric" of the field. Stone stripes, downfield curvature, long-axis alignment of boulders downfield, and small lobate masses of talus toward the bottom of slopes beyond the boulder field are all evidence suggesting that the rocks have moved. Trees, however, show that movement has not occurred recently. The formation of boulder fields and boulder colluvium on ridge flanks in the central Appalachians is associated with the periglacial climate of the Wisconsinan (Pleistocene) glaciation, when freeze and thaw cycles were much more extreme and breakup of the ridge tops and downslope movement were active. River of Rocks (site 12) and Blue Rocks (site 14) are similar deposits, the discussion of which pertains to this site.

25. CREST OF PETERS MOUNTAIN AND
26. POWELLS VALLEY OVERLOOK

These two overlooks are located about one mile from each other. Peters Mountain Overlook provides a view to the south. Third Mountain marks the axis (center) of the syncline that terminates to the west. Hard, resistant sandstone of the Pocono Formation underlies Second Mountain and Peters Mountain, whereas the valleys adjacent to the mountains are composed of soft, easily eroded shale and siltstone.

The view from Powells Valley Overlook is to the north across a broad valley. Agricultural land in this valley is on red shale and sandstone (Catskill Formation) of Devonian age. Berry Mountain, underlain by hard, resistant sandstone (Pocono Formation), may be seen in the distance. A water gap cut through Berry Mountain by the Susquehanna River is also visible.

As shown on the sketch map, the Trail traverses the north limb of a large syncline all the way from the DeHart Reservoir Overlook (site 24) to Cove Mountain (site 29). The sketch map outlines the relationship between the folded ridges, the twisting river and streams, and the route of the Trail.

View looking south from Peters Mountain Overlook.

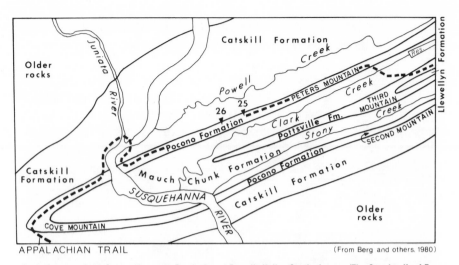

(From Berg and others. 1980)

Geologic map of the Peters Mountain Overlook and Powells Valley Overlook area. (The Spechty Kopf Formation is included in the Pocono Formation in this drawing.)

On the crest of Peters Mountain, the Trail passes over the steeply dipping sandstone (Pocono Formation) of Shikellimy Rocks.

View looking north from Powells Valley Overlook.

The rocks underlying the ridges are predominantly layers of hard sandstone and conglomerate; they are very resistant to erosion and hence form high mountains. The valleys, on the other hand, are composed of soft, easily eroded beds of shale and siltstone.

27. SUSQUEHANNA RIVER OVERLOOK

Between Powells Valley Overlook and the Susquehanna River, the Trail passes many high, steeply south dipping outcrops of sandstone and conglomerate layers and often traverses rock rubble along the spine of the ridge. The effects of periglacial weathering, expressed along this mountain crest, are as well demonstrated here as anywhere in Pennsylvania. A few hundred feet west of this overlook a massive section of the ridge-crest outcrop has slumped to the north and slid part of the way down the mountain flank.

During the Wisconsinan glaciation, which reached a maximum extent in this region about 15,000 years ago, glacial ice did not come in contact with Peters Mountain. However, as the ice front advanced and retreated northward, intense freeze-thaw conditions occurred for a long time, accelerating rock weathering along upturned ridge edges,

South-dipping crossbedded sandstone and conglomerate near the crest of Peters Mountain.

A close look at the quartz-pebble conglomerate of the Pocono Formation on Peters Mountain. This hard rock, along with quartzite, underlies the ridges that make up the route of the Appalachian Trail in this area.

Sandstone and conglomerate boulder rubble (Pocono Formation) on the crest of Peters Mountain between Powells Valley Overlook and the Susquehanna River. Note the blazes marking the route over the rocky terrain.

including Peters Mountain, far to the south of the glacial border. Water in fractures froze, thawed, and froze again, causing powerful ice wedging and natural quarrying of the ridge crest, and deposition of large quantities of boulders along ridge flanks. This type of weathering by mass wasting of rock continues today, but at a much slower rate than it did in the periglacial climate. So, it was periglacial conditions that caused the loose, rocky terrain, serrated bedrock topography, boulder colluvium, and boulder fields over which much of the Trail passes in Pennsylvania. The Appalachian Trail hiker has a better feel for the results of this periglacial activity than do most visitors to the state.

The Susquehanna River Overlook is outstanding. The pronounced westward bend in the river and the abrupt, steep termination of Peters Mountain on the inside of the bend combines with the junction between the Susquehanna and Juniata Rivers to provide a magnificent view of a varied landscape.

View to the north from the Susquehanna River Overlook.

The view to the north is over red sandstone, siltstone, shale, and mudstone of the Catskill Formation, one of the most widespread rock units in central and eastern Pennsylvania.

Looking to the north, the confluence of the rivers is in interbedded layers of mudstone, siltstone, and very fine to medium-grained sandstone. Beyond the second bridge on the Juniata River are alternating thin beds of sandstone, siltstone, and shale. These rocks are less resis-

View to the south from Peters Mountain near the Susquehanna River Overlook. The Great Valley may be seen in the distance through a "window" created by the water gaps.

tant to erosion than the formations that underlie the mountains and, as a result, the core of the anticline here is weathered more deeply than its flanks.

Looking to the south, the Susquehanna passes through two water gaps and into the Great Valley.

Mineral resources are extracted from the Susquehanna at Haldeman Island just upstream from its junction with the Juniata. Sand and gravel is quarried from unconsolidated river and stream sediments to be used for construction aggregate and roadway material. Also, there is a coal dredge in the vicinity of Amity Hall. River coal is carried here by the Susquehanna River from the coal fields to the north. The river flows down from New York State, passing through the Northern Anthracite field in Pennsylvania on its way south. Over the years, coal-mining activities and coal breaker and washing operations deposited significant amounts of finely ground coal over long stretches of the river. This has been profitably extracted from the river bed beyond the coal fields for many years, there being enough in one case to supply the Holtwood power generating station farther downstream.

River ice has frequently caused a problem at the confluence of these two long rivers. In the spring, ice flowing down the Susquehanna can block the flow of ice from the smaller Juniata. The resultant high build-up of ice in the lower few miles of the Juniata has been responsible for stripping the shore of trees and cottages.

28. HAWK ROCK

Hawk Rock, which is below the crest of Cove Mountain, is a ledge of sandstone (Catskill Formation). It derives its name from the hawk migrations that may be observed from this rock.

Here, the rock layers dip steeply to the south. Sandstone and conglomerate (Pocono Formation) form the ridge crest above. The ridge adjacent to Cove Mountain on the north is Pine Ridge, underlain by hard, reddish-brown sandstone (Catskill Formation) that is resistant to weathering and forms the ridge here as well as the prominent bench on the north flank of Peters Mountain across the Susquehanna. Between the Susquehanna River Overlook (site 27) and the Susquehanna River, the Trail passes over good exposures of this sandstone formation along the steep edge of the ridge part of the way down the mountain.

Hawk Rock, a ledge of sandstone (Catskill Formation) high on the north flank of Cove Mountain.

29. COVE MOUNTAIN

The geology of Cove Mountain may be illustrated best by a geologic block diagram. In the diagram one can see that the Mauch Chunk Formation has been squeezed between non-yielding limbs of the Pocono Formation. Millions of years after this folding, diabase intruded the fractured rocks of the syncline.

From this location it is easy to relate the surface topography to the rock type and geologic structure. To the southwest, Pine Ridge is a resistant sandstone in the Catskill Formation, the same rock unit as that cut by Sherman Creek below Hawk Rock. It wraps around the cove, forming a subsidiary ridge on the south flank of the syncline between this location and Pa. Route 850. To the northeast, the valley bottom is on the Mauch Chunk Formation, a sequence of brick-red shale, siltstone, and sandstone layers which were squeezed, folded, and faulted in the grip of the surrounding layers of hard, resistant conglomerate and sandstone of the Pocono Formation.

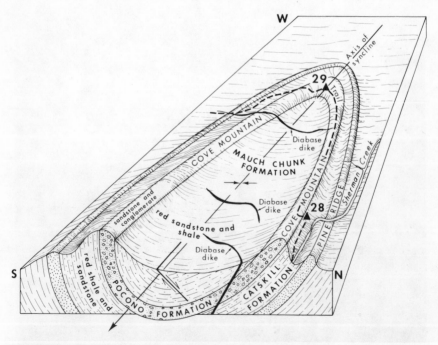

Block diagram of Cove Mountain. (The Spechty Kopf Formation is included in the Pocono Formation in this drawing.)

Quartz-pebble conglomerate in the Pocono Formation, the resistant rock that forms Cove Mountain, is frequently encountered between the Susquehanna River and this site. It is particularly well exposed along the rock wall remaining from excavation for a pipeline at this location.

Diabase is an intrusive igneous rock emplaced along fractures and zones of weakness in the crustal rocks from an igneous melt at depth. It is encountered at the foot of the mountain south of Cove Mountain, on the east side of the Trail just into the woods from a Pennsylvania Game Commission parking area and at the junction of Pa. Route 850 and Miller Gap Road. The rock can be recognized by its rusty-appearing exterior and dense, dark-gray interior.

Between Cove Mountain and Blue Mountain the Trail crosses narrow valleys and ridges adjacent to the Great Valley. Blue Mountain is the prominent first mountain delineating the Appalachian Mountain section from the Great Valley section of the Valley and Ridge physiographic province.

30. GREAT VALLEY

The view south from Blue Mountain is across the Great Valley. A gently rolling topography is characteristic of this valley, as it is underlain by relatively easily eroded limestone, dolomite, and shale. Limestone and dolomite (carbonate rocks) are soluble from precipitation over long periods of time.

This overlying gentle topography masks an intensely folded and faulted subsurface. Thrust faults and overturned folds are common. The Great Valley represents one of the most geologically complex areas in the Commonwealth. The beautiful rural countryside does not reflect the twisted and contorted rocks beneath.

Many limestone springs may be found in the Great Valley, but just west of the Trail is Boiling Springs, the seventh largest in Pennsylvania (11,500 gallons per minute) and the most picturesque in the Great Valley.

The Great Valley as seen from Blue Mountain has a gently rolling landscape. South Mountain is in the background.

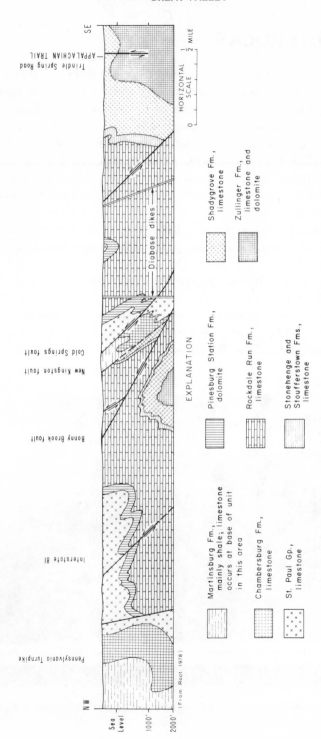

Geologic cross section of folded and faulted limestone, dolomite, and shale underlying the Great Valley between Blue Mountain and South Mountain.

31. WHITE ROCKS

A point of outstanding demarcation, White Rocks stands at the northern end of the Blue Ridge physiographic province, a series of mountains that dominate the route of the Appalachian Trail southward through Maryland, Virginia, Tennessee, North Carolina, and Georgia, ending at Mount Oglethorpe, the original southern terminus of the Trail. The Blue Ridge province in Pennsylvania is represented by South Mountain, an anticlinorium composed of ancient rocks. The Trail at White Rocks and southward is along the northwestern flank of this anticlinorium.

White Rocks is composed of a light-gray, hard, vitreous quartzite (Antietam Formation, Cambrian age). It is a fractured, blocky, ridge-forming rock.

Clay, iron, and some manganese have been extracted commercially near the base of the mountain in the vicinity of White Rocks. White clay, used as filler in the manufacture of paper, in brick making, and in the production of hydraulic white cement, is associated with the limestone and dolomite of the Tomstown Formation near its contact with

Thick beds of light-gray, vitreous quartzite make up White Rocks, a prominent exposure on the flank of South Mountain.

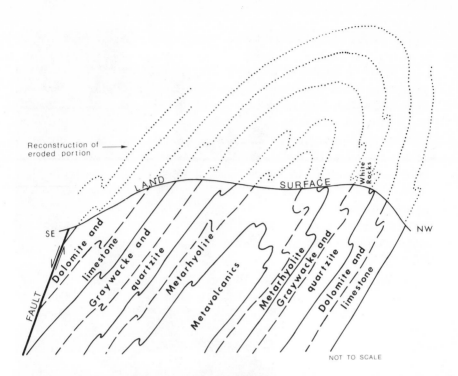

Diagrammatic cross section of South Mountain. This type of geologic structure is called an anticlinorium, a regional anticline that has smaller folds on each of its limbs.

the quartzite of the Antietam Formation. Clay deposits are thought to have formed by hot solutions associated with geologic fault zones and by chemical weathering of the limestone and dolomite in the Tomstown Formation (Hosterman, 1968).

In a similar geologic environment at the foot of the steep slope of South Mountain, not far above the valley floor, iron and manganese ore deposits have been worked in this area. Mining was started prior to 1750. The iron and manganese originate from weathering and leaching of limestone and dolomite. Thick accumulations of clay, which resulted from weathering of the carbonates (limestone and dolomite), formed the host material into which iron and manganese oxide minerals were deposited by replacement.

There is not much lateral continuity to these deposits, and for this reason exploration and mining were sporadic in the late nineteenth and first half of the twentieth centuries. Many of the overgrown pits

Quartzite of the Antietam Formation at White Rocks, on the north limb of the South Mountain anticlinorium.

and excavations along the Trail on the mountain slope north of White Rocks are the result of past exploration efforts.

32. POLE STEEPLE

Pole Steeple, 0.3 mile north of the Appalachian Trail and 2 miles east of Pine Grove Furnace, is not visible from the Trail but is definitely worth a visit. This pillar of quartzite rises over South Mountain as its name implies. Fractures in the surrounding rock formations and major faults nearby account for the steeple standing out as it does, rather than forming a continuous rock ledge along the flank of Piney Mountain. The formation of rock fractures and weakening along boundary fault zones allowed progressive erosion to isolate the steeple as a remnant on the mountainside.

The rock at Pole Steeple is hard, light-gray, vitreous quartzite (Montalto Member of the Harpers Formation). Metarhyolite and dolomite are less resistant rocks in the valley to the northwest in the vicinity of Laurel Forge Pond. These two rock types were faulted upward against the quartzite and subsequently eroded more rapidly and deeply than the quartzite so that they now occupy a lower topographic position.

Pole Steeple.

33. PINE GROVE FURNACE

Any discussion of the geology of the Appalachian Trail deals with a description of the events that occurred in the geologic past, resulting in the rocks and landforms that characterize the Trail environment today. Also included are the influences of the works of man, past and present, to extract mineral resources: limestone and slate near Allentown, coal northeast of Harrisburg, and iron in the Great Valley and in South Mountain. The Pine Grove area is an excellent example of the latter.

Pine Grove was an active iron-mining area and furnace for 120 years, from 1773 to 1893. It was one of many iron ore deposits mined in eastern Pennsylvania. In total, Pennsylvania's iron industry since the early 1700's had a notable effect upon the economy of this state and played an important wartime role from the Revolutionary War to the Civil War.

Remains of the Pine Grove Furnace. Furnaces of this type were generally built against a small hill to facilitate placing the charge of iron ore, charcoal, and limestone. Limestone was added to separate the impurities (slag) from the molten iron.

Iron at Pine Grove was extracted from open pits and shallow underground workings, smelted in a furnace, and then shipped by wagon and later by railroad to nearby forges and foundries. There it was further processed and cast for cannon shot, stoves, pipe, tools, and wrought iron. South Mountain iron ore yielded iron that was satisfactory for casting in foundries but was not sufficiently malleable for forging until it was processed with higher quality "valley" ore such as that found at Dillsburg, Cornwall, and French Creek.

During most of the history of mining in South Mountain, potential reserves of ore were not considered. It was plentiful, as was limestone for flux and wood for charcoal. A blue haze of smoke covered the hills from hundreds of charcoal-making hearths. Timber was cut and burned day and night. Approximately 24 cords of wood were required to produce the charcoal to smelt enough ore to yield just one ton of pig iron.

The history of the charcoal-iron industry in South Mountain is well documented in the hiker's *Guide to the Appalachian Trail from the Susquehanna River to the Shenandoah National Park* (Potomac Appalachian Trail Club, 1970). The geology and origin of the ore bodies, however, were not studied until the mines had been in operation for many years. There had always been enough ore, and not until it became scarce did people wonder about its occurrence and genesis. General descriptions of the South Mountain ore banks and tables of chemical analyses were published in 1846 in the First Pennsylvania Geological Survey reports. After that, in 1886, the Second Pennsylvania Geological Survey reports gave detailed descriptions of the shape, size, and attitude of these ore deposits, but, as exemplified by the following from the Pennsylvania Geological Survey's Annual Report of 1886 (d'Invilliers, 1887, p. 1467), it was already too late.

> For, inasmuch as this opening has practically been abandoned for the time being, many of the working faces and old shafts were filled in and it was with difficulty that several of them were located at all.

Deposits of iron ore throughout South Mountain are near the contact between the dolomite (Tomstown Formation) and the quartzite (Antietam Formation), and are often associated with faulting. The most notable deposits along the Appalachian Trail are near White Rocks, Pine Grove, and Quarry Gap. Pine Grove is the best example, showing remains of a 90-foot-deep ore pit (part of which is now Fuller Lake), the furnace, and the iron master's mansion house, now a hostel for hikers. South of Fuller Lake, across a small ridge, other ore pits are visible.

The mansion at Pine Grove Furnace is built on limestone and dolomite of the Cambrian Tomstown Formation.

The deposition of these iron deposits is due to leaching by the downward movement of slightly acidic groundwater. Iron mineralization came primarily from widely scattered grains of magnetite in the rocks of the mountain (Hosterman, 1968). Iron precipitated from solution and concentrated into slightly sandy beds of dolomite as the acidity of the solution changed in the presence of the carbonate rock. Fault zones and developing solution openings in the dolomite provided sites of deposition for the iron minerals. This slow chemical process was active over a long period of geologic time as the dolomite weathered from the South Mountain anticlinorium to its present position.

Close to the turn of the century this charcoal-iron industry which had been so prominent in southeastern Pennsylvania came to a close. Extensive iron ore deposits of the Lake Superior region were being developed, and the Bessemer process, invented in 1855, was introduced to the United States, giving the world its first low-priced steel. In addition, the supply of wood for charcoal making was all but gone and, at the end, the deep pit at Pine Grove was flooded.

34. SUNSET ROCKS

Sunset Rocks on the crest of Little Rocky Ridge is a striking geologic feature. Quartzite and conglomerate are the primary rock types found here. Of particular note, however, are the white quartz bands and lenses in this rock exposure. Also small folds, synclines and anticlines, may be found. Although seen in miniature, the South Mountain anticlinorium (the regional geologic fold) would look like these folds, but would be approximately 10 miles in width.

Along the crest of Little Rocky Ridge a portion of Sunset Rocks is balanced as an erosion remnant of quartzite and conglomerate.

White quartz bands and lenses at Sunset Rocks.

Small folds, synclines and anticlines, at Sunset Rocks.

35. LEWIS ROCKS

Lewis Rocks, approximately 0.7 mile southeast of the Trail, is a magnificent rock peak which affords a view of South Mountain to the east. The rocks are coarse-grained conglomerate and quartzite (Weverton Formation of Cambrian age). Widely spaced white quartz stringers, blobs, and layers may also be found in these rock outcrops.

Close examination of Lewis Rocks also reveals fractures that have iron-stained linings and raised edges. These fractures appear as narrow channelways, the walls of which are harder than the surrounding rock. This hardening of the fracture walls and the small, resistant ridges along the fracture edges are due to precipitation of iron from iron-rich solutions passing through the fractures. These solutions impregnate the rock to a depth of an inch or more along the opening, hardening that portion and causing it to stand as a narrow lip after rock weathering.

Two small, connected potholes are located in a ravine below Lewis Rocks. The origin of this geologic feature is interesting. Tumbling Run

Lewis Rocks, named after Lewis the Robber, is at the head of a steep ravine above Dead Woman Hollow. The rock is coarse-grained conglomerate and quartzite (Weverton Formation).

Small potholes in the streambed below Lewis Rocks. These potholes originally formed near the toe of a waterfall which has now migrated several tens of feet upstream.

carries, especially during storms, hard rock fragments that cut the stream channel by mechanical abrasion. Waterfalls are present in this channel, and often at the base of a falls small rock fragments get caught in current eddies and the abrasive grinding action of the rock fragments cut circular holes (potholes) in the streambed. Many times rounded pebbles and cobbles may be found in the bottom of a pothole. These pebbles represent the "drills" which, combined with the fast flowing water, drill the hole.

36. BIG FLAT TOWER

Big Flat fire tower stands on top of Big Flat Ridge at an elevation of 2,060 feet. The tower rises an additional 53 feet above ground level and the upper platform provides a spectacular view of the Great Valley, here called the Cumberland Valley, to the north and west. Anyone climbing the tower should use caution. When wet, the steps and decks are slippery, and the wind can be strong near the top.

In addition to the view of the Cumberland Valley, the hiker will see the rounded and irregularly dissected hilly topography of South Mountain.

About 3.5 miles northwest of the Trail in Hairy Springs Hollow is the well-preserved Big Pond Iron Furnace stack. The geology of the Big Pond area and the origin of the iron ore are the same as at Pine Grove Furnace (site 33).

View of the Cumberland Valley from Big Flat Tower. Blue Mountain is in the distance.

Big Pond Iron Furnace stack. This well-preserved structure is in the valley about 3.5 miles from Big Flat Tower. It was active during approximately the same period as Pine Grove Furnace.

37. QUARRY GAP AND CALEDONIA PARK

At Quarry Gap, look beyond the thick rhododendrons and notice the narrow notch through which the Trail passes in Quarry Hill. Also note the steep-sided ridge itself and the change in topography as the Trail goes south toward Caledonia Park and across U.S. Route 30. Old iron ore pits exist at the gap, and the furnace remains at the Caledonia Ironworks, once owned by Thaddeus Stevens and significant to the Civil War, may also be seen here. Other relics of past activity adjacent to the Trail are circular charcoal hearths from nineteenth-century charcoal preparation.

Iron ore was extracted from numerous pits along the southeast base of Quarry Hill at the contact between quartzite (Antietam Formation) and dolomite (Tomstown Formation). The geologic setting of these iron ore deposits is similar to that at Pine Grove Furnace (site 33). For

The iron furnace at Caledonia Ironworks, built in 1837, was converted to a monument to Thaddeus Stevens, who owned the enterprise until his death in 1868.

a discussion of the overall history of this area the reader is referred to *Guide to the Appalachian Trail from the Susquehanna River to the Shenandoah National Park*, published by the Potomac Appalachian Trail Club (1970).

The geology in the vicinity of Caledonia Park is the most complex in all of the South Mountain anticlinorium. This is primarily due to

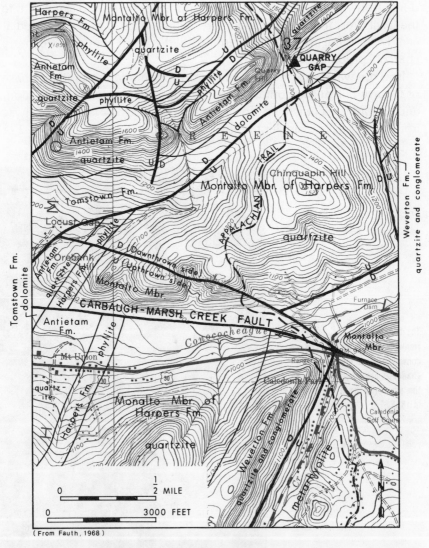

(From Fauth, 1968)

Geologic sketch map of the vicinity of Caledonia State Park. Note the large number of geologic faults along which bedrock segments have been rearranged like pieces of a three-dimensional puzzle.

numerous faults, breaks in the earth's crust, which have rearranged rock units like pieces of a three-dimensional puzzle. There are eleven known faults. The largest and most spectacular is called the Carbaugh-Marsh Creek fault. It starts near Cashtown northwest of Gettysburg, passes through South Mountain, offsetting a portion of the entire range, crosses the Trail at Caledonia Park, turns in the Great Valley, and extends south to Martinsburg, West Virginia. Apparent movement along the fault has been measured as about 2-1/2 miles in the South Mountain area. It has topographic expression in the orientation of Conococheague Creek, the course of which was determined by the trace of the fault. At Caledonia Park the creek turns abruptly to flow westward in a transverse valley developed on the fault trace (the line of intersection between a fault and the ground surface). The straight, wide creek valley and the offset ridges are all evidence of the existence of this fault that you will be able to see from the Trail. Hundreds of millions of years ago, during a major period of mountain building, South Mountain was structurally deformed. No movement has occurred along the fault for hundreds of millions of years, since the time of mountain building, and there is no reason to believe that any movement will occur in the near future.

38. SNOWY MOUNTAIN TOWER

A panoramic view of South Mountain is seen from this tower. To the north, the hiker is looking across a valley underlain by very old volcanic rocks (metarhyolite). These rocks are some of the oldest to be found in the Commonwealth. The buildings of the State Sanatorium in South Mountain Village are on these volcanics.

Facing south, the view is over a dissected upland of quartzite. Snowy Mountain Tower is also on this quartzite (Weverton Formation), as are many of the high points along the Trail in South Mountain. The quartzite is a ridge former because it is hard and highly resistant to erosion, whereas some of the volcanic rocks are softer and weather rapidly, forming valleys.

The view southeast from Snowy Mountain Tower is over a dissected upland underlain by quartzite and across a valley underlain by volcanic rocks. Green Ridge is in the background.

The view to the north from Snowy Mountain Tower is across a valley underlain by Precambrian volcanic rocks. The ridge in the background is Chinquapin Hill near Caledonia Park; the trace of the Carbaugh-Marsh Creek fault is in front of the ridge.

39. CHIMNEY ROCKS AND MONUMENT (SHAFFERS) ROCK

Chimney Rocks and Monument Rock are close to one another and are combined here and discussed as one location. Both are excellent rock outcrops and vistas. Monument Rock is one of the most picturesque places, geologically speaking, in all of the South Mountain region. It is called Shaffers Rock locally and in trail guidebooks, but is named Monument Rock on the Iron Springs 7-1/2-minute topographic map.

Chimney Rocks is an overlook above the valley containing the Waynesboro Reservoir. The reservoir and valley are underlain by metarhyolite and quartzite. Chimney Rocks is light-gray coarse-grained quartzite of the Weverton Formation.

While standing on top of Monument Rock or looking up at it from below, the majesty of the rocks that make the mountains is strongly felt. Near the toe of boulders and talus accumulated from weathering of the cliffs above, the Potomac Appalachian Trail Club's Hermitage Cabin fits this scene nicely.

A large fracture (joint) opening in the quartzite at Monument Rock appears as a narrow street in a "city" of rocks.

Chimney Rocks, composed of quartzite (Weverton Formation), resembles a large chimney ravaged by time.

Monument Rock, composed of quartzite (Weverton Formation), stands as a monolith above Hermitage Cabin and the surrounding terrain.

A joint opening in Monument Rock is spectacular from the top or base.

40. BUZZARDS ROOST

Buzzards Roost, near the Appalachian Trail, is related to a quartzite quarry. Few rock exposures other than this exist in the area, so this small quarry is an important site from which to study the geology of this section of South Mountain.

Buzzards Roost east of the Appalachian Trail near Beartown. A quartzite quarry provides an ideal roost for local buzzards.

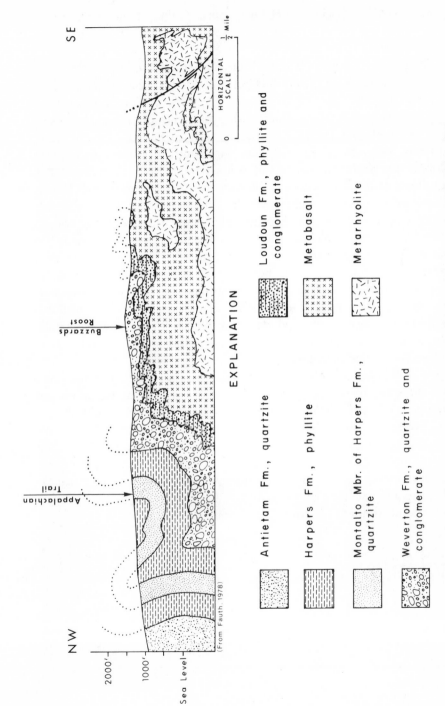

Geologic cross section from northwest to southeast across the west limb of the South Mountain anticlinorium near Buzzards Roost.

Folding of the rocks is prominent here; faulting takes a secondary role. There are only two short faults south of Beartown and Buzzards Roost adjacent to Pa. Route 16.

The mining history of South Mountain has involved iron ore deposits concentrated near the contact between the quartzite (Antietam Formation) and the dolomite (Tomstown Formation). The origin and history of these mines in the nineteenth century are discussed in the section on Pine Grove Furnace (site 33).

There was also some prospecting in South Mountain for copper. Copper prospects closest to the Appalachian Trail are in the valley southeast of the Trail between Monument Rock and Buzzards Roost. There are about a dozen inactive pits, all associated with old volcanic rocks. Copper occurs mainly in its native state along with the copper minerals cuprite, malachite, azurite, chrysocolla, and chalcocite. In this area there are short tunnels, shafts, and drifts as evidence of past prospecting, but none have been exploited commercially recently. Copper occurs as fracture fillings in the rocks. Pennsylvania Geological Survey Atlas 129c (Fauth, 1978) contains a discussion of these copper deposits and mapped locations of each. Provided that permission is obtained from landowners, some of the sites may make good mineral-collecting localities.

Industrial minerals have been mined in South Mountain in the past, and presently there are several active quarries. Crushed rock for aggregate is extracted from quartzite of the Weverton Formation, Antietam Formation, and Harpers Formation. Metabasalt provides beautifully colored material for roof granules now being taken from a large quarry near Charmian and, at Buzzards Roost, a quartzite quarry that is intermittently active provides material for stained roof granules. Metarhyolite has been quarried and pulverized for use as a filler material in paint, rubber, and roofing compounds.

CONCLUSION

And so at Pen Mar, as the Trail leaves Pennsylvania, our story ends. We hope you have enjoyed the geology of the Trail and have acquired an appreciation of the role geology plays with respect to the scenery and the route of the Trail. Hiking the Trail is always a good experience; with this geologic background we hope that the experience is enhanced.

REFERENCES

Becher, A. E., and Root, S. I. (1981), *Groundwater and geology of the Cumberland Valley, Cumberland County, Pennsylvania*, Pennsylvania Geological Survey, 4th ser., Water Resource Report 50, 95 p.

Berg, T. M., and Dodge, C. M. (1981), *Atlas of preliminary geologic quadrangle maps of Pennsylvania*, Pennsylvania Geological Survey, 4th ser., Map 61, 636 p.

Berg, T. M., Edmunds, W. E., Geyer, A. R., and others (1980), *Geologic map of Pennsylvania*, Pennsylvania Geological Survey, 4th ser., Map 1, scale 1:250,000, 3 sheets.

Bloom, A. L. (1978), *Geomorphology: a systematic analysis of Late Cenozoic landforms*, Englewood Cliffs, New Jersey, Prentice-Hall, Inc., 510 p.

Bucek, M. F. (1971), *Surficial geology of the East Stroudsburg 7-1/2 minute quadrangle, Monroe County, Pennsylvania*, Pennsylvania Geological Survey, 4th ser., Atlas 214c, 40 p.

Crowl, G. H., and Sevon, W. D. (1980), *Glacial border deposits of late Wisconsinan age in northeastern Pennsylvania*, Pennsylvania Geological Survey, 4th ser., General Geology Report 71, 68 p.

d'Invilliers, E. V. (1887), *Report on the iron ore mines and limestone quarries of the Cumberland-Lebanon Valley*, in *Annual report of the Geological Survey for 1886*, Pennsylvania Geological Survey, 2nd ser., Annual Report 1886, Pt. IV, p. 1409-1567.

Drake, A. A., Jr. (1969), *Precambrian and Lower Paleozoic geology of the Delaware Valley, New Jersey-Pennsylvania*, in Subitzky, Seymour, ed., *Geology of selected areas in New Jersey and eastern Pennsylvania and guidebook of excursions*, New Brunswick, N.J., Rutgers University Press, p. 51-131.

Epstein, J. B. (1966), *Structural control of wind gaps and water gaps and of stream capture in the Stroudsburg area, Pennsylvania and New Jersey*, U.S. Geological Survey Professional Paper 550-B, p. B80-B86.

_____ (1973), *Geologic map of the Stroudsburg quadrangle, Pennsylvania-New Jersey*, U.S. Geological Survey Geologic Quadrangle Map GQ-1047, 3 p., 1 plate, scale 1:24,000.

Epstein, J. B., and Epstein, A. G. (1967), *Geology in the region of the Delaware to Lehigh Water Gaps*, Annual Field Conference of Pennsylvania Geologists, 32nd, East Stroudsburg, Pa., 1967, Guidebook, 89 p.

_____ (1972), *The Shawangunk Formation (Upper Ordovician(?) to Middle Silurian) in eastern Pennsylvania*, U.S. Geological Survey Professional Paper 744, 45 p.

Epstein, J. B., Sevon, W. D., and Glaeser, J. D. (1974), *Geology and mineral resources of the Lehighton and Palmerton quadrangles, Carbon and Northampton Counties, Pennsylvania*, Pennsylvania Geological Survey, 4th ser., Atlas 195cd, 460 p.

Fauth, J. L. (1968), *Geology of the Caledonia Park quadrangle area, South Mountain, Pennsylvania*, Pennsylvania Geological Survey, 4th ser., Atlas 129a, 133 p.

_____ (1978), *Geology and mineral resources of the Iron Springs area, Adams and Franklin Counties, Pennsylvania*, Pennsylvania Geological Survey, 4th ser., Atlas 129c, 72 p.

Foose, R. M. (1945), *Iron-manganese ore deposit at White Rocks, Cumberland County, Pennsylvania*, Pennsylvania Geological Survey, 4th ser., Mineral Resource Report 26, 35 p.

Frazer, Persifor, Jr. (1877), *Report of progress in the counties of York, Adams, Cumberland and Franklin*, Pennsylvania Geological Survey, 2nd ser., Report CC, p. 240-262.

Freedman, Jacob (1967), *Geology of a portion of the Mount Holly Springs quadrangle, Adams and Cumberland Counties, Pennsylvania*, Pennsylvania Geological Survey, 4th ser., Progress Report 169, 66 p.

Geyer, A. R. (1969), *Hickory Run State Park — Hickory Run Boulder Field*, Pennsylvania Geological Survey, 4th ser., Park Guide 2.

Geyer, A. R., Smith, R. C., II, and Barnes, J. H. (1976), *Mineral collecting in Pennsylvania*, 4th ed., Pennsylvania Geological Survey, 4th ser., General Geology Report 33, 260 p.

Hoskins, D. M. (1969), *Fossil collecting in Pennsylvania*, Pennsylvania Geological Survey, 4th ser., General Geology Report 40, 126 p.

_____ (1976), *Geology and mineral resources of the Millersburg 15-minute quadrangle, Dauphin, Juniata, Northumberland, Perry, and Snyder Counties, Pennsylvania*, Pennsylvania Geological Survey, 4th ser., Atlas 146, 38 p.

Hosterman, J. W. (1968), *White clay deposits near Mt. Holly Springs, Cumberland County, Pennsylvania*, in Cloos, Ernst, Freedman, Jacob, Hole, Gilbert, and others, *The geology of mineral deposits in south-central Pennsylvania*, Annual Field Conference of Pennsylvania Geologists, 33rd, Harrisburg, Pa., 1968, Guidebook, p. 38-51.

Lapham, D. M., and Bassett, W. A. (1964), *K-Ar dating of rocks and tectonic events in the Piedmont of southeastern Pennsylvania*, Geological Society of America Bulletin, v. 75, p. 661-667.

Lash, G. G. (1978), *The structure and stratigraphy of the Pen Argyl Member of the Martinsburg Formation in Lehigh and Berks Counties, Pennsylvania*, unpublished M.S. thesis, Lehigh University, Bethlehem, Pa.

Martin, R. A. (1971), *Geology of the Devil's Racecourse boulderfield, Dauphin County, Pennsylvania*, unpublished M.Ed. thesis, Millersville State College, Millersville, Pa.

McPhee, John (1981), *Basin and range*, New York, Farrar, Straus, and Giroux, 216 p.

Miller, B. L., Fraser, D. M., and Miller, R. L. (1939), *Northampton County, Pennsylvania*, Pennsylvania Geological Survey, 4th ser., County Report 48, 496 p.

Miller, B. L., Fraser, D. M., Miller, R. L., and others (1941), *Lehigh County, Pennsylvania*, Pennsylvania Geological Survey, 4th ser., County Report 39, 492 p.

Peele, Robert, editor (1927), *Mining engineers' handbook*, v. 1, New York, John Wiley and Sons, p. 1212.

Pennsylvania Geological Survey (1981), *Glacial deposits of Pennsylvania*, Pennsylvania Geological Survey, 4th ser., Map 59.

Potomac Appalachian Trail Club (1970), *Guide to the Appalachian Trail from the Susquehanna River to Shenandoah National Park*, Washington, D.C., Potomac Appalachian Trail Club, 7th ed., p. 19-23.

Potter, Noel, Jr., and Moss, J. H. (1968), *Origin of the Blue Rocks block field and adjacent deposits, Berks County, Pennsylvania*, Geological Society of America Bulletin, v. 79, p. 255-262.

Root, S. I. (1968), *Geology and mineral resources of southeastern Franklin County, Pennsylvania*, Pennsylvania Geological Survey, 4th ser., Atlas 119cd, 118 p.

_____ (1976), *Engineering problems in areas of limestone springs*, Pennsylvania Geology, v. 7, no. 2, p. 6-9.

_____ (1978), *Geology and mineral resources of the Carlisle and Mechanicsburg quadrangles, Cumberland County, Pennsylvania*, Pennsylvania Geological Survey, 4th ser., Atlas 138ab, scale 1:24,000.

Royer, D. W. (1981), *Caledonia and Pine Grove Furnace State Parks, Cumberland, Adams, and Franklin Counties—Geologic features and iron ore industry*, Pennsylvania Geological Survey, 4th ser., Park Guide 15.

_____ (1982), *Swatara State Park, Lebanon and Schuylkill Counties—Geologic features and fossil sites*, Pennsylvania Geological Survey, 4th ser., Park Guide 16.

Rupp, I. D. (1845), *History of Northampton, Lehigh, Monroe, Carbon and Schuylkill Counties*, Lancaster, Pa., G. Hills, 554 p. [reprinted by Arnold Press, N.Y., 1971, 568 p.].

Smith, R. C., II (1977), *Zinc and lead occurrences in Pennsylvania*, Pennsylvania Geological Survey, 4th ser., Mineral Resource Report 72, 318 p.

Willard, Bradford (1962), *Pennsylvania geology summarized*, Pennsylvania Geological Survey, 4th ser., Educational Series 4, 17 p.

Wood, G. H., Jr. (1968), *Geologic map of the Tower City quadrangle, Schuylkill, Dauphin, and Lebanon Counties, Pennsylvania*, U.S. Geological Survey Geologic Quadrangle Map GQ-698, scale 1:24,000.

Wood, G. H., Jr., and Kehn, T. M. (1961), *Sweet Arrow fault, east-central Pennsylvania*, American Association of Petroleum Geologists Bulletin, v. 45, p. 256-263.

_____ (1968a), *Geologic map of the Pine Grove quadrangle, Schuylkill, Lebanon, and Berks Counties, Pennsylvania*, U.S. Geological Survey Geologic Quadrangle Map GQ-691, scale 1:24,000.

_____ (1968b), *Geologic map of the Swatara Hill quadrangle, Schuylkill and Berks Counties, Pennsylvania*, U.S. Geological Survey Geologic Quadrangle Map GQ-689, scale 1:24,000.

GLOSSARY

Allochthonous. Pertaining to a stratigraphic sequence of bedrock that has been moved from its original site of origin by tectonic forces. This term is used in the explanation on the geologic map (Plate 1).

Anticline. A fold of rock strata that is convex upward.

Autochthonous. Pertaining to a stratigraphic sequence of bedrock formed or occurring in place where found. This term is used in the explanation on the geologic map (Plate 1).

Colluvium. A deposit of loose, heterogeneous rock fragments and soil material deposited by mass wasting along a steep slope.

Diabase. An intrusive igneous rock, the main components of which are the minerals labradorite and pyroxene.

Dip. The angle that a planar rock surface (e.g., a bedding plane) makes with the horizontal; measured perpendicular to the *strike*.

Fining-upward cycle. A general vertical decrease in grain size, ranging from conglomerate (pebble-sized particles) to shale (mud-sized particles) in a sedimentary rock sequence; caused by a decrease in current velocity during deposition. This term is used in the explanation on the geologic map (Plate 1).

Foliation. A planar arrangement of structural features, such as cleavage, caused by deformation of a bedrock sequence.

Graywacke. Hard, well-indurated, poorly sorted, generally coarse grained sandstone of variable composition.

Intrusive. Pertaining to the emplacement of igneous magma into pre-existing rock.

Metarhyolite. A metamorphosed, extrusive igneous rock that was originally rhyolite.

Metasedimentary. Pertaining to a sedimentary rock that has been changed by metamorphism. Its original characteristics have been changed by heat and/or pressure during burial under other rock sequences in the geologic past.

Periglacial. Extreme and variable cold climatic conditions at the margin of a continental glacier which promote frost action and accelerated mass-wasting processes.

Phyllite. An altered mudstone formed by regional metamorphism, intermediate in metamorphic intensity between slate and mica schist; characterized by pronounced cleavage due to uniform orientation of flaky minerals.

Plunge. The inclination of a fold axis measured by its departure from the horizontal.

Slickenside. A polished and smoothly striated surface that results from friction along a fault plane.

Strike. The direction that a rock surface takes as it intersects the horizontal.

Syncline. A fold of rock strata that is concave upward.

Talus. Rock fragments of any size lying against the base of a cliff and derived from weathering of the cliff.

Trace. The intersection of a fault surface with the ground surface.

Wildflysch. A mappable stratigraphic unit displaying large and irregularly sorted blocks and boulders resulting from tectonic fragmentation. This term is used in the explanation on the geologic map (Plate 1).

STRATIGRAPHIC COLUMN OF GEOLOGIC UNITS CROSSED BY THE APPALACHIAN TRAIL IN PENNSYLVANIA

PERIOD OF GEOLOGIC TIME	AGE (millions of years)	SYMBOL ON GEOLOGIC MAP (Plate 1)			NAME OF FORMATION (Thickness)	GEOLOGIC DESCRIPTION
TRIASSIC	240 to 205 m.y. ago	℞d			Diabase (10 to 60 feet)	Medium-gray, fine- to coarse-grained diabase consisting of labradorite (feldspar) and augite (a pyroxene).
PENNSYLVANIAN	330 to 290 m.y. ago	ℙl			Llewellyn Formation (3,500 feet)	Interbedded brown to light-gray, medium- to coarse-grained sandstone and conglomerate with numerous anthracite coal beds and dark-gray to black shales.
PENNSYLVANIAN	330 to 290 m.y. ago	ℙp			Pottsville Group (1,400± feet)	Light- to dark-gray, fine- to coarse-grained sandstone and conglomerate with some minable coals; subordinate amounts of siltstone and shale.
MISSISSIPPIAN	360 to 330 m.y. ago	Mmc			Mauch Chunk Formation (4,500± feet)	Red to gray and greenish-gray shale and fine- to medium-grained, flaggy sandstone with siltstone and claystone.
MISSISSIPPIAN	360 to 330 m.y. ago	Mp			Pocono Formation (1,720± feet)	Light- to dark-gray, fine- to coarse-grained sandstone and gray quartz-pebble conglomerate. Contains olive-gray to dark-gray shale and siltstone interbeds and coal stringers.
MISSISSIPPIAN	360 to 330 m.y. ago	MDsk			Spechty Kopf Formation (570± feet)	Light- to olive-gray, fine- to medium-grained sandstone with olive-gray to dark-gray shale and siltstone. Grayish-red shale near top and conglomerate at base and in middle in some areas.
DEVONIAN	410 to 360 m.y. ago	Dck	Dcd	Catskill Formation	Duncannon Member (1,400± feet)	Light-olive-gray to olive-gray, fine- to medium-grained, crossbedded, micaceous sandstone with grayish-red siltstone and claystone.
DEVONIAN	410 to 360 m.y. ago	Dck	Dccf	Catskill Formation	Clarks Ferry Member (225 feet)	Grayish-purple to brownish-gray and light-gray to olive-gray, medium- to coarse-grained, micaceous, crossbedded sandstone, conglomeratic sandstone, and conglomerate with thin interbeds of dark-gray shaly claystone.
DEVONIAN	410 to 360 m.y. ago	Dck	Dcsc	Catskill Formation	Sherman Creek Member (3,100 feet)	Grayish-red to brownish-gray and light-gray to olive-gray, very fine to fine-grained sandstone, siltstone, and chippy-weathering claystone.
DEVONIAN	410 to 360 m.y. ago	Dck	Dciv	Catskill Formation	Irish Valley Member (2,350 feet)	Interbedded red and nonred claystone, shaly siltstone, and sandstone arranged in repetitive sequences of marine and nonmarine units.

STRATIGRAPHIC COLUMN (Continued)

PERIOD OF GEOLOGIC TIME	AGE (millions of years)	SYMBOL ON GEOLOGIC MAP (Plate 1)			NAME OF FORMATION (Thickness)	GEOLOGIC DESCRIPTION
DEVONIAN	410 to 360 m.y. ago			Dtr	Trimmers Rock Formation (1,840 feet)	Medium-gray to olive-gray siltstone and very fine grained, sandy siltstone; planar bedded; individual beds in outcrop extend long distances without appreciable change in thickness. Many layers are laminated.
		Dh	Hamilton Group	Dmh Dm	Mahantango and Marcellus Formations (2,100 feet)	Mahantango—brown to olive shale with interbedded sandstones which are dominant in places (Montebello Sandstone Member). Includes brown sandstone (Turkey Ridge Member) in parts of central Pennsylvania. Fossiliferous in upper part. Marcellus—black, fissile, carbonaceous shale. Tioga bentonite at base in eastern Pennsylvania.
				Doo	Onondaga and Old Port Formations (200 feet)	Onondaga—medium- to dark-gray calcareous shale and argillaceous limestone. Old Port—light-gray, coarse-grained sandstone, siliceous sandstone, chert, shale, and limestone.
				DSop	Onondaga Formation through Poxono Island Formation (1,200 ± feet)	Gray calcareous, sandy shale; brown sandstone; interbedded limestone, dolomite, and shale; gray calcareous sandstone; gray shaly limestone and olive-gray calcareous shale, siltstone, and sandstone.
SILURIAN	435 to 410 m.y. ago			Swc	Wills Creek Formation (650 feet)	Gray to yellowish-green, interbedded calcareous shale, siltstone, and sandstone; some shaly limestone and dolomite.
		Sbm		Sb	Bloomsburg Formation (450 feet)	Grayish-red claystone, argillaceous siltstone, and very fine to fine-grained sandstone.
					Mifflintown Formation (300 feet)	Interbedded dark-gray shale and medium-gray fossiliferous limestone.
				Sc	Clinton Group (1,000 ± feet)	Predominantly Rose Hill Formation —light-gray to brownish-gray, fossiliferous shale and dark-reddish-gray, very fine to coarse-grained, ferruginous sandstone. Includes Keefer Formation—light- to dark-gray, fossiliferous sandstone.

STRATIGRAPHIC COLUMN *(Continued)*

PERIOD OF GEOLOGIC TIME	AGE (millions of years)	SYMBOL ON GEOLOGIC MAP (Plate 1)	NAME OF FORMATION (Thickness)	GEOLOGIC DESCRIPTION
SILURIAN	435 to 410 m.y. ago	St	Tuscarora Formation (550 to 700 feet)	Light-gray and light-olive-gray, very fine to medium-grained, tightly cemented orthoquartzite with some medium-gray and medium-olive-gray silty shale.
		Ss	Shawangunk Formation (2,000± feet)	Light-to dark-gray, fine- to very coarse grained quartzite and conglomerate containing thin shale interbeds. Includes four members. Equivalent to Tuscarora and Rose Hill Formations to the west.
ORDOVICIAN	500 to 435 m.y. ago	Ojb	Juniata and Bald Eagle Formations (1,830± feet)	Grayish-red siltstone, shale, and very fine to medium-grained sandstone with gray to olive-gray, fine- to coarse-grained sandstone.
		Om	Martinsburg Formation (Thickness unknown; structurally distorted; may be up to 12,800 feet thick)	Buff-weathering, dark-gray shale, silty in places, with thin interbeds of siltstone and fine-grained graywacke. Thin-bedded, platy-weathering, dark-gray argillaceous limestone and bands of shale in the lower portion.
		Oc	Chambersburg Formation (770± feet)	Gray argillaceous limestone with shale partings.
		Oh	Hamburg sequence (Estimated at 3,000 feet; structurally deformed)	Allochthonous (transported) rocks of the Hamburg overthrust, composed of greenish-gray phyllitic shale containing zones of graywacke sandstone and limestone.
		Osp	Saint Paul Group (up to 1,000 feet)	Finely crystalline "birdseye" limestone containing black chert and dolomite. Fossiliferous.
		Ops	Pinesburg Station Formation (300± feet)	Light-gray, thick-bedded dolomite and some limestone.
		Orr	Rockdale Run Formation (2,000± feet)	Light-gray, laminated, fine-grained limestone containing some chert and dolomite.
		Os	Stonehenge Formation (up to 1,500 feet)	Fine-grained limestone; coarse-grained, conglomeratic limestone at base.
CAMBRIAN	570 to 500 m.y. ago	€sg	Shadygrove Formation (up to 1,000 feet)	Light-gray pure limestone containing pink limestone and cream-colored chert.
		€z	Zullinger Formation (2,500± feet)	Medium-gray, thin- to thick-bedded, interbedded limestone and dolomite.

STRATIGRAPHIC COLUMN *(Continued)*

PERIOD OF GEOLOGIC TIME	AGE (millions of years)	SYMBOL ON GEOLOGIC MAP (Plate 1)	NAME OF FORMATION (Thickness)	GEOLOGIC DESCRIPTION
CAMBRIAN	570 to 500 m.y. ago	€e	Elbrook Formation (3,000± feet)	Light-gray calcareous shale and silty limestone with medium-gray limestone and dolomite.
		€wb	Waynesboro Formation (up to 1,000 feet)	Interbedded red to purple shale and sandstone; some dolomite and limestone interbeds.
		€t	Tomstown Formation (up to 1,000 feet)	Massive dolomite containing thin shaly interbeds.
		€a	Antietam Formation (300± feet)	Gray, buff-weathering quartzite.
		€h €hm	Harpers Formation Montalto Member (1,500± feet)	Greenish-gray phyllite and schist and gray quartzite.
		€wl	Weverton and Loudoun Formations (1,300± feet)	Gray quartzite and quartzose conglomerate; some slate and metamorphosed sandstone.
*		mb (Catoctin volcanics)	Metabasalt (Estimated at 1,000 feet)	Dark-gray, fine-grained intrusive igneous rock.
		mr (Catoctin volcanics)	Metarhyolite (Estimated at 1,000 feet)	Light-gray to purple, schistose intrusive igneous rock.
		vs (Catoctin volcanics)	Greenstone schist (100± feet)	Fine- to medium-grained, light- to medium-green schist; includes probable metavolcanic rocks.

*Zircon minerals from metarhyolite indicate that the Catoctin volcanics are 820 million years old.

GEOLOGIC TIME CHART

SUBDIVISIONS				Age estimates of boundaries, in million years
Phanerozoic	Cenozoic Era	Quaternary Period	Present day, 1983	0
			Holocene Epoch	0.010 (10,000 yrs.)
			Pleistocene Epoch	2
		Tertiary Period	Pliocene Epoch	5
			Miocene Epoch	24
			Oligocene Epoch	38
			Eocene Epoch	55
			Paleocene Epoch	63
	Mesozoic Era	Cretaceous Period	Late Cretaceous Epoch	96
			Early Cretaceous Epoch	138
		Jurassic Period		205
		Triassic Period		⌒240
	Paleozoic Era	Permian Period		290
		Carboniferous Periods	Pennsylvanian Period	⌒330
			Mississippian Period	360
		Devonian Period		410
		Silurian Period		435
		Ordovician Period		500
		Cambrian Period		⌒570
Proterozoic	Precambrian			2,500
Archeozoic			Oldest known rocks in U.S.	3,600

INDEXES

ALPHABETICAL LISTING OF
GEOLOGIC SITES ALONG THE APPALACHIAN TRAIL
(Site numbers are in parentheses)

LISTING OF GEOLOGIC SITES BY TOPOGRAPHIC MAP

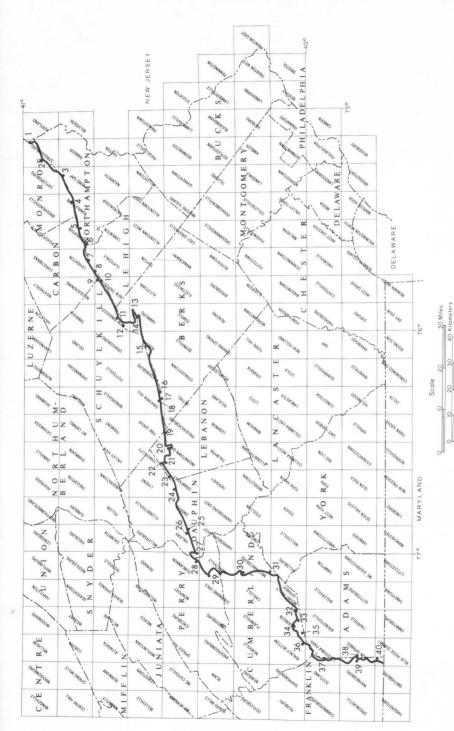

LOCATION MAP OF GEOLOGIC SITES ALONG THE APPALACHIAN TRAIL